Dutch Elm Disease – The Early Papers

Selected Works of Seven Dutch Women Phytopathologists

Translated and Prepared by
Francis W. Holmes and Hans M. Heybroek

APS PRESS
The American Phytopathological Society

Acknowledgment of Financial Contributors

We are grateful to the following firms and societies for contributing toward the costs of publication of this Phytopathological Classic:

Campbell Scientific Co, Logan, Utah

Canada Chapter of International Society of Arboriculture

F A Bartlett Tree Expert Co, Stamford, Connecticut

Forestry Suppliers, Jackson, Mississippi

Foundation Fonds Landbouw Export Bureau, Wageningen, The Netherlands

International Society of Arboriculture, Urbana, Illinois

Memorial Research Trust of International Society of Arboriculture, Urbana, Illinois

New England Chapter of International Society of Arboriculture

Princeton Nurseries, Princeton, New Jersey

Rhône-Poulenc Co, Research Triangle Park, North Carolina

Rohm & Haas Co, Philadelphia, Pennsylvania

Shell International Petroleum Co Ltd, London, England

Stauffer Chemical Co, Dayton, New Jersey

Westchester County Tree Protection Association, Yonkers, New York

---

Cover: Schematic presentation of a diseased tree
(from "An Unknown Disease Among the Elms, II,"
by Dina Spierenburg; see page 34). Drawing by W. H. Ruisch.

Library of Congress Catalog Card Number: 90-82571
International Standard Book Number: 0-89054-110-8

©1990 by The American Phytopathological Society

Printed in the United States of America

The American Phytopathological Society
3340 Pilot Knob Road
St. Paul, Minnesota 55121, USA

Contents

Remarks by the Translators
of This Phytopathological Classic

We read these papers with an enormous hindsight, developed in 70 years of DED outbreaks, and through thousands of scientific publications on the disease. It is fascinating to look through the eyes of these young women afresh at a "new" disease, and to follow how they reached their (mostly correct) conclusions.

Translations were by F. W. Holmes, except for the major portion of chapter 3, which was by H. M. Heybroek. Dates of translations are given in footnotes in each chapter. Comments or sentences in brackets anywhere in the book are those of the translator. Boldface page numbers in brackets indicate the top of a page in the original work.

Seven Dutch Women Scientists Whose Early Research Is Basic to Our Knowledge of the "Dutch Elm Disease"

By Francis W. Holmes

In this Phytopathological Classic, our Society presents an often-overlooked fact: that the foundation of our knowledge of the Dutch elm disease (DED) was laid by women scientists. Six of those represented in this translation performed the earliest research on this disease, and the seventh summarized it. All worked in The Netherlands.

At the very outset, let us realize that the name "Dutch elm disease" should be seen as a compliment, a term of honor, recognizing early and thorough Dutch research on this important disease.

Considering the centuries of commercial and naval rivalry between the Dutch and the English, it may have been natural to assume that "Dutch elm disease" was a negative English usage like "Dutch courage," "Dutch treat," "Dutch uncle," or "a Dutchman's chance." This assumption, however, is fallacious. The real origin of the name comes from the fact that all the early reports and studies on this disease came from The Netherlands.

When the disease was first noticed in 1919, it had already covered most of Belgium and the Netherlands, plus parts of northern France. Many Dutch, French, and German writers called it just the "elm disease" or "elm death" (*de iepenziekte; la maladie des ormes; das Ulmensterben*). Only the Dutch, however, gave it much serious scientific study. A long series of useful, fundamental, scientific publications appeared in technical journals, all by Dutch authors.

It was natural, then, that when the disease reached England in 1926, it became known there as the "Dutch" elm disease.

However, the very first use of the name "Dutch" in print, a year earlier, apparently was in Germany (Brussoff, A., 1925, Die hollandische Ulmen-krankheit...eine Bakteriosis, Centralblatt für Bakteriologie 63:256).

The name Dutch elm disease is now accepted even by Dutch scientists—especially when writing in English—who called it that only a few years later (Buisman, 1932, Tijdschift der Nederlandse Heidemaatschappij 40(10):338-344; see translation later in this Classic). After all, there are a good many other elm diseases, so it's not very helpful to call this one just "elm disease." We felt free, then, in this Classic, to call it the Dutch elm disease and even to use the common acronym DED.

In 1919, Ms. **Barendina Gerarda Spierenburg** started work at the Dutch Phytopathologischen Dienst (PD) in the town of Wageningen, near the Dutch Agricultural University. At that time, the PD was a combination of a national extension service, plant disease clinic, and plant quarantine service; now it has a very comprehensive task, including matters relating to the Pesticides Act and interaction with other members of the European Economic Community. Dina Spierenburg was the first to report (in 1921) the excessive dying off of elm trees in the 1919 and 1920 growing seasons. The disease apparently arose (or perhaps arrived) while everyone was distracted by the devastation and danger of World War I. In 1920, DED clearly had been around for at least two or three years, because some still-standing, large, dead elm trees had already lost bark and twigs. Most people assumed that the outbeak had begun about 1917 or 1918.

Naturally the first question was: What is killing the elms? Many suggestions were made. In France, drought and/or war-gas poisoning were proposed. In Belgium, the mycologist Biourge considered this a variation of the Nectria canker disease. In Germany, Brussoff isolated a bacterium that he named *Micrococcus ulmi* and that he thought infested the soil. (But no one ever seems to have found *M. ulmi* again!) A proper, thorough, sober, scientific, work was undertaken at the Phytopathologisch Laboratorium Willie Commelin Scholten (WCS), which had been founded in Amsterdam in 1895 and moved to Baarn in 1921.

One of the earliest laboratories in the world to be devoted wholly to plant diseases, the WCS came to be associated with three universities: the University of Utrecht, the University of Amsterdam, and the Free University of Amsterdam.

Dr. Ritzema Bos directed the WCS Laboratory from 1895 to 1906 and then left for the Agricultural University at Wageningen. The second Director (from 1906 to 1950) was Prof. Dr. **Johanna Westerdijk**. She had the creative idea of funding DED studies by raising money from the 996 separate municipalities of The Netherlands. She assigned study of the cause of DED to one of her graduate students in 1920, and later she assigned other personnel to DED: to confirm the causal agent, start resistance research, investigate fungal physiology, and the like. She herself made many public speeches describing progress, advocating further research, and passionately defending the discoveries of her staff. She often spoke to lay audiences, to commodity associations of the agricultural industries, and especially to the Dutch Heath Society (i.e., the Dutch Heathland Reclamation Society).

Marie Beatrice Schwarz was, at that time, preparing a doctoral dissertation (on the pathology of peach and willow tree diseases) under Prof. Westerdijk. We can imagine what she thought when one day her professor told her she must also include the new problem with elm trees. But she loyally added

2

the elm trees to the others. Today, she is famous for her work on elms—to the point that almost no one knows what it was that she learned about peaches or willows!

Schwarz isolated microbes from the diseased elms and inoculated healthy elms with the microorganisms she had obtained. She concluded that one of them—a certain fungus—caused the new disease. She named its imperfect stage (the stage bearing nonsexual spores, which was all that she had found) *Graphium ulmi* Schwarz, new species. By the rules of botanical nomenclature, this name still may be used if one is dealing with only the imperfect stage, as is so often the case with DED. Dr. Schwarz soon went to the Dutch East Indies, where she was employed and where, a few years later, she married T. C. Schol, becoming Mrs. Dr. M. B. Schol-Schwarz.

In the meantime, although Dr. Schol-Schwarz had indeed discovered the true cause of DED, very few people in the plant disease research world believed her. It was, after all, a rare thing for a graduate student to discover the cause of a new disease, especially one that would become world-famous. Many a doctoral degree is given for far less useful discoveries over small points of detail. Apart from her colleagues at the WCS Laboratory, only five others—Prof. Dr. H. W. Wollenweber at the German National Experiment Station in the Dahlem suburb of Berlin; two women scientists in Germany (Prof. Dr. Countess von Linden and Lydia Zenneck, both at the University of Bonn); and two scientists in England, again one a woman ("Malcolm Wilson and Dr. Mary Wilson," as Westerdijk put it)—could perceive that Bea Schwarz was right.

All through the 1920s, this controversy raged, while the disease continued to spread through Europe and warnings began to be sounded in America that it might someday arrive there. So Prof. Westerdijk assigned another of her students to conduct a further study of the cause. Young Dr. **Christine Johanna Buisman** had just completed a doctoral thesis on root rot of calla lilies, but she spent the rest of her short life studying nothing but DED!

Prof. Westerdijk asked Dr. Buisman to repeat and expand Schwarz's earlier elm study, so as either to confirm or to refute the findings. Buisman embarked at once into an elaborate program of isolations from and inoculations into elm trees at many different times of the year. She promptly found that Schwarz had been right: the fungus *Graphium ulmi* did cause DED.

The reason why Schwarz never could produce real wilt (and why others also had had trouble repeating Schwarz's work), Buisman discovered, was that the period when elms are susceptible to artificial inoculations is limited to about 6 weeks in early summer: June and part of July in the Dutch climate. In addition, freshly potted or transplanted trees will rarely show symptoms upon inoculation. Buisman succeeded, using the same kinds of elm as Schwarz, when she inoculated at the right time of year and in well-established trees.

In 1931, when Dr. Schol-Schwarz and her husband returned to The Netherlands for a leave of absence from work in the East Indies, the fun-loving scientific community at the Willie Commelin Scholten Laboratory planned a little surprise for her. No one had ever told Dr. Schol-Schwarz that her scientific findings had been widely doubted by most European scientists. So her former colleagues said to her, "Come along with us next week, Bea, to the (annual) plant disease conference in Ghent (Belgium)". She went, and she was sitting in the audience when the results of her graduate studies were triumphantly vindicated in a paper presented by Dr. Christine Buisman.

Meanwhile Prof. Westerdijk did not rest content to have only Schwarz and Buisman studying this disease. She assigned another doctoral candidate, **Maria Sara Johanna Ledeboer**, to study the physiology of the fungus: What acidity did it prefer? What elements did it need for nutrition? What materials were toxic to it? Dr. Ledeboer, in her dissertation in 1934, opened for the first time the question of possible injections into elm of fungicidal materials against the Dutch elm disease fungus. She squarely faced and discussed a problem that still faces us: the phytotoxicity of such chemotherapeutants to the very elm trees they are to protect!

But Dr. Buisman was not limiting her studies to the cause of the disease. In 1929 or 1930 (not all references agree), in conjunction with planning by Prof. Westerdijk, she initiated a far-sighted and far-reaching program to search for disease resistance in all elm species and elm populations and to enhance the resistance by breeding. She gathered elms from every available source and selected among them by inoculation.

However it was left for Dr. **Johanna Catharina Went**, who took the torch from Buisman's hands in 1936, to do the first crossings of various (more-or-less resistant) elms with one another, creating hybrids by interbreeding. This program still continues after 55 years. To date, nine named, DED-resistant elm clones have been released to the public through the nursery industry, to replace the centuries-old clone *Ulmus hollandica* 'Belgica', the Belgian elm that formerly grew by the millions along roads and dikes in The Netherlands.

This research has encountered both expected and unexpected difficulties. The Dutch knew that they must combine DED resistance with wind resistance and still retain a beautiful tree form. Salt-laden North Sea winds are a part of everyday life in The Netherlands—and the Dutch public has an appreciative eye for a well-formed tree, especially for trees used in intact, symmetrical rows along straight roads.

No one foresaw the fact that clone number 24, found by the Dutch in Spain and issued in 1936 as the Christine Buisman elm, even though highly DED resistant, would prove almost useless to The Netherlands itself. At that latitude (indeed the same latitude as Labrador) it dies back severely from the coral spot canker disease, caused by another fungus, *Nectria cinnabarina* Fr. In addition, its shape and level of wind resistance were unsatisfactory. But this elm clone *is* a suitable tree in Spain, in Italy, and in the United States at latitudes where the patterns of day-length variation more closely resemble those of its Iberian origin. (This is true despite the fact that in such U.S. areas the winters are far colder than in either Spain or The Netherlands.)

4

Even this elaborate resistance-breeding program was not enough for Dr. Christine Buisman's energy and capabilities. Meanwhile, she discovered the perfect, or sexual, fruiting stage of *Graphium ulmi*, first in the laboratory and then in nature. She entered the allied field of mycology to describe it and named it *Ceratostomella ulmi* Buisman.

Ceratostomella seemed then to be the logically correct group, but it suffered from an error that had been made a half-century earlier, in 1881, by a man named Winter. Winter had combined two different types of related fungi under the name *Ceratostomella*. Many fungi of both types were put into this genus in the 51 years between 1881 and 1932. Meanwhile there were many failures in the attempts to find a correct genus name for the non-*Ceratostomella* parts of this mixture. Indeed, it was not until 1952 that Buisman's name for the sexual stage of the DED fungus was changed to what some scientists now consider the correct name: *Ceratocystis ulmi* (Buisman) C. Moreau. Other scientists, however—especially those in Europe—now prefer *Ophiostoma ulmi* (Buisman) Nannfeldt. By the botanical rules, Buisman's own name is still attached to either of these Latin binomials, in honor of her initial discovery.

Dr. Buisman's discovery opened a whole field of research into inheritance of variations in this fungus species, a favorite subject of mine. Dr. Buisman showed that two identical-looking but internally different isolates of the fungus had to mate to form the sexual fruiting bodies. She called these two groups "+" and "−." Today they are called "A" and "B." These are now known to be, not two sexes but two mating or compatibility groups. In nature, each of them contains only hermaphroditic cultures (cultures able to act as either male or female), which can mate only with cultures of the opposite mating group. But in a laboratory, male-only cultures have been derived in both A and B—through the loss of the ability to act as females. (As yet, no female-only cultures have been found.)

Dr. Buisman's other articles consist of long, thoughtful annual reports on the DED research being performed. In only eight years (between her doctorate in 1927 and her tragic death after a cancer operation in 1935), this remarkable scientist, whose name was posthumously assigned to the most widely used of the DED-resistant elms she eventually discovered, published 33 technical scientific papers in five languages: Dutch, English, French, German, and Italian. She spent a year in Italy as consultant (DED was killing elm trees that were used as supports for grape vines in a major industry there) and another year in the United States, mostly in Massachusetts at Radcliffe College and at the Arnold Arboretum of Harvard University. Somehow she also found time to teach a mycology course in Portugal! She was also unofficially active in helping a young entomologist at the Agricultural University in Wageningen, J. J. Fransen, to complete his doctoral dissertation, in which he showed that elm bark beetles could carry the DED fungus from one elm tree to another. Perhaps Buisman's tremendous efforts tired her too much. Anyway, to the sorrow of many, including myself, she seems not to have been strong enough

5

in 1935 to survive a necessary operation (by some reports a minor operation, by other reports an operation for cancer).

Of Buisman's 33 technical papers, I have translated 32 (plus two memoria about her) into English (the other was already in English). A complete bibliography of her publications can be obtained from me at cost. Funds are still sought to publish the whole collection as a booklet in her honor. She was very far-seeing in her work. Had Buisman lived, DED research might by now be much farther advanced than it is. Despite the some 3,500 papers that have been published since the disease arose (an average of about one a week for the past 65 years!), there is still much to learn.

At once Dr. Johanna C. Went took over management of the valuable program of selection and breeding for DED resistance. She carried this program through the dangerous and difficult occupation years of World War II, still under direction of Prof. Dr. Johanna Westerdijk. Dr. Went ran that program from 1935 to 1954, a total of two decades, and then turned to other research, delving into processes of decomposition of leaf litter of an oak forest floor.

During the World War II occupation, food, shoes, clothes, fuel, and research materials were scarce. Survival itself was doubtful. Work had to be done under continual German surveillance. It was an effort even to force oneself to go to work each day, to say nothing of giving close attention to rigorous scientific disciplines like phytopathology or the genetics of disease resistance.

We all owe a debt for our present-day resistant elms to the courage of the scientists of the Phytopathological Laboratory Willie Commelin Scholten and also to those in the companion Centraalbureau voor Schimmelcultures (CBS), the Dutch Central Agency for Fungus Cultures, then housed in the same building in the town of Baarn, who continued their research despite the German occupation.

The CBS had been created, and was directed, by Prof. Westerdijk. It now rates as one of the three principal world collections of fungi in culture, the other two being the Commonwealth Mycological Institute, Kew Gardens, England, and the National Type Culture Collection, Beltsville, Maryland, USA.

Phytopathology owes a special debt to Dr. Westerdijk, who not only founded a dynasty of able and productive plant pathologists (about 50% women and 50% men) and laid a sound basis for understanding and controlling the devastating DED, but also founded a world-renowned fungus culture collection. Her persistence during the war and the survival of her research program in the difficult aftermath provide much to admire. In 1950, after 44 years as director of both WSC and CBS, she finally had to surrender the reins of control of both of the famous institutes she had so long led. She died in 1953.

The new director of the Phytopathological Laboratory Willie Commelin Scholten was Prof. Dr. **Louise Catharina Petronella Kerling**, who had survived World War II in a Japanese concentration camp. (The new director of the Centraalbureau voor Schimmelcultures was Prof. Dr. Agathe L. van Beverwijk. Both were women scientists of repute.)

Prof. Kerling continued to encourage DED research and even increased it. At one point (1962–1963), she had three researchers working on it at once— a whole "DED corridor" in the Lab! In addition, she herself contributed to DED literature, with a publication on its history. She had the insight, too, to appoint a plant physiologist to study DED in her laboratory. Prof. Kerling retired in 1972 and died in 1985, ending the famous and productive body of early work on DED by Dutch women plant pathologists over the half century 1919–1972. Their work was so well done and is so highly regarded that Dutch research on DED continues to hold a leading position in the world.

The fourth director of the Phytopathological Laboratory Willie Commelin Scholten (1972–1985) was Prof. Dr. Koen Verhoeff, and the fifth (1985–) is Dr. Bob Schippers. The WCS Laboratory thus had two women as directors (for a total of 72 years) and three men (for a total of 27 years). In 1987 came the sad news that the WCS Laboratory is scheduled to be closed in 1991, for lack of funding cooperation among universities. For generations of plant pathologists the world around, it has long been a professional Mecca.

After her husband's death under the Japanese occupation, Dr. Marie Schol-Schwarz returned to the CBS at Baarn. I was honored and delighted there in 1963 to do a small study with her on an atypical variant in the coremial stage of the fungus *Graphium ulmi*—the fungus that she herself had named 42 years earlier. Dr. Schol-Schwarz died in 1969, in Baarn.

In this Phytopathological Classic, the spotlight is on these seven women. That is perhaps not entirely fair: even in the 1920s some men also made contributions to DED research, working with Westerdijk. For example, S. Broekhuizen, whose dissertation at Leiden University in 1929 was entitled *Wondreaksies van hout* (Wound Reactions of Wood), included studies of the brown discoloration that characterizes xylem tissue of DED-infected elms. J. J. Fransen (as mentioned above) demonstrated the transmission of this disease fungus by elm bark beetles. Cooperation with men was generally good, as the testified by the reports of the Committee for the Study and Control of the Elm Disease (an independent nationwide group set up to gather funds, authorize research, and publish reports).

But from the research foundation laid by these women phytopathologists grew the huge structure of study that has led to our present knowledge of DED. Others who publish new discoveries on DED probably cite their publications more often than any other references in the scientific literature.

Those in the biological sciences who now seek proper recognition and support for women desiring phytopathological careers should take courage from these famous examples—just as those in the physical and chemical sciences look with pride to the accomplishments and fame of, for example, Marie Curie in discovering radium. But the large role of women throughout the history of plant disease research is little known even to plant pathologists, and it is far less known to today's public. It deserves greater recognition and high respect.

Dina Spierenburg in 1959, on the occasion of her being honored at the 60th anniversary celebration of the Plantenziektenkundige Dienst, in Wageningen.

CHAPTER 2

Barendina Gerarda Spierenburg

1880–1967

Barendina Gerarda Spierenburg, known to her colleagues as "Dina" or "Din," was born in Rotterdam on April 10, 1880. After several years of teaching, she returned at age 35 to her own academic advancement. She studied biology from 1915 to 1919 at the University of Utrecht, in a teachers' training program (M.O. = *Middelbare Onderwijs*), meanwhile assisting a Dr. Wickham and teaching in a Rotterdam high school.

In 1919, Ms. Spierenburg was appointed to the Institute for Phytopathology (the name of which almost at once changed to Phytopathological Service) in Wageningen, led by Prof. Johan Ritzema Bos. This Institute was the forerunner of five present-day institutions: three University departments (Phytopathology, Virology, and Nematology); the Plant Pathology Research Institute (IPO); and the Plant Protection Service (PD), which specialized then in diagnosis, extension, and quarantine and more recently has undertaken many other responsibilities, some of them evolving from the Pesticides Act.

In 1919, upon the formation of this last unit (renamed in 1921 the Plantenziektenkundige Dienst or PD), Spierenburg—together with Messrs. H. Maarschalk and T. A. C. Schroevers—became one of the original staff. She was phytopathologist under Ir. N. Van Poeteren for 26 years until, upon her retirement in 1945, she was succeeded by Ms. Dr. A. Jaarsveld. Especially in those formative years, the PD was noted for its comradely, pioneer spirit.

Spierenburg's duties were many, as specimens of all kinds of plant diseases, in a never-ending stream, flowed across her desk and laboratory counters for diagnosis and advice. Almost from the outset she was immersed in "the unknown disease of elms," now famous as Dutch elm disease (DED). Her very professional studies paralleled and corroborated those of Marie Beatrice Schwarz in Baarn, although the two laboratories, with their divergent tasks, had rather little communication.

Spierenburg was the first to point out to phytopathologists that there was an epidemic among the elms. Her 1921 map of widespread DED distribution, accompanied by her report of isolations from early wood layers of street elms in Renkum (1912–1913) and Rotterdam (1913–1915) show us that DED was present in Europe before World War I—indeed probably from at least as early as about 1900–1905—and that its origin there could have had nothing to do with the war gases or Chinese labor battalions that were present from 1914 to 1918.

Spierenburg clearly recognized the correct cause of DED, although she preferred to wait to see *all* symptoms (including wilt) develop in the trees she had inoculated with her "fungus mixture" of *Cephalosporium* sp. and *Graphium* sp. As Westerdijk said in 1929, if Spierenburg had only tried single-spore cultures, she would at once have realized that these were just different spore stages of a single fungal species! There was no mean-spirited clash as to credit for priority of discovery: in an article in 1930, Spierenburg explicitly credited Schwarz as having first demonstrated the true cause of DED.

Spierenburg could give only partial attention to the elm problem because of her broad responsibilities at the PD throughout her 26-year career. In the 1930s, she published important studies on diseases of cabbage and gladiolus. She was noted for her realistic approach and for her sense of humor, which sustained her in the face of difficulties and disappointments. In retirement she actively studied languages and music and traveled to Finland, Lapland, Greece, Turkey, Russia, France, Spain, and Tangiers.

On December 10, 1967, after a sickness of only a few days, Dina Spierenburg died in Rotterdam, the city of her birth. She was 87 years, 8 months old.

A memorial was published in the *Netherlands Journal of Plant Pathology* in 1968 (74:33-34).

An Unknown Disease Among the Elms

Spierenburg, Dina
1921. *Een onbekende ziekte in de iepen.* Plantenziektenkundige Waarnemingen I.
Mededeling 18 van den Phytopathologischen Dienst (januari): 3-10.
[Plant Disease Observation I. Communication 18 of the Phytopathological Service]*

[p. 3] One of the diseases which has provided a whole lot of work this past year for the officials of the Phytopathological Service, is a disease in the elms, which, so far as we know, was not observed earlier in the country [The Netherlands]. Since as yet we know only a little about this disease, I can report here only how the disease looks, what has become known to us about the disease from letters or orally, and that which the research has yielded to date.

In January of the year 1920 we got some elm twigs, submitted from Hoeven near Oudenbosch, about which an accompanying note read that the twigs originated from trees which the previous year had also supplied the Service [with] material for research: in September 1919 Mr. ONRUST, Technical Officer 1st Class at the Phytopathological Service, who sent us the twigs, had sent from Hoeven young elm trees [probably trees less than 3 m tall], the leaves of whose young shoots then suddenly died. A bit later, in the month of September of the same year, we received from Tilburg from the Chief of Municipal Works there, Mr. F. KRUGERS, some material of diseased elms which suddenly withered there, too. In both cases, thus as early as 1919, we cultured various fungi out of the diseased elms, and reported about this to the people involved. We did not give much further attention to the disease.

In 1920 that would be entirely different. From January until late in the fall, there came from all parts of the nation mailings and complaints about a disease among the elms, with the request to do everything to save the threatened trees.

Because of intense pressure of work at the laboratory during the summer of 1920, [it was] October [when] we first had a chance to get to know, on the spot [in situ], the disease which suddenly had arisen in various parts of our land.

The Inspector, Head of the Phytopathological Service, Mr. N. VAN POETEREN, visited Tilburg, Mr. T.A.C. SCHROEVERS, phytopathologist of the Service, went to Venlo and Zaltbommel, [p. 4] while I visited Rotterdam, Schiedam, Delft, Oud-Beierland and Opheusden myself. In this last place are the nurseries where in most cases the young elms are bought for the cities. In the conversations preliminary to these visits, we quickly noticed that in the various places we were dealing with the same phenomenon. Venlo could be an exception to this, perhaps also Oud-Beierland.

*[Translated by F. W. Holmes, 2/1985]

Even though our visit was quite late in the year, nevertheless, because of the prolonged, beautiful autumn weather, the trees still bore enough foliage to show the syndrome to us distinctly. Besides, I found later, on my visit to Opheusden, the diseased individuals still can be easily recognized among trees entirely without leaves.

The disease appearance [syndrome]. With trees in foliage one sees in the top a wholly dried out and shrivelled-up mass of dead leaves and branches, in the midst of the remaining, still living portions of the tree. The twigs are dry, shrivelled, parched as it were; the smaller twigs, moreover, are recurved at the tip. In most cases, small, dried-up buds for the following year are present.

The green leaves on diseased trees give the impression, in comparison with leaves of healthy individuals, as though the foliage were somewhat dry and brittle; they don't look so fresh. Often one finds a brown, dried margin around the otherwise still green leaf.

Where, here and there, dormant buds have sprouted out right on the trunk, and thus small, one-year-old twigs are present, these turn out to be almost all dead, while the dried-out foliage is still on them and the tip of the twig is curved back here, too.

In cross section, branches and trunks show, in the wood close to the bark, a ring of small, brown flecks; sometimes, a bit nearer to the center, a second or third ring. See plate I, fig. 1. Occasionally it happens that the entire cut surface of a thinner twig appears strewn with such tiny, brown specks. These little spots extend to the tip of the outermost branches. With thicker branches, the discoloration is mostly limited to the most recent annual rings, while the older annual rings appear normal. It does sometimes occur with a thick branch, that the interior wood is for the most part brown-colored. Although the central part of the healthy elm trunk is often of a darker hue, yet we can clearly see in the cases in question that we [p. 5] are dealing with an infiltrate [infusion], such as often occurs with woody plants whose roots entirely or partially have turned rotten.[1] The color of the infiltrate is [a] lighter brown than the brown stipples of the discolored rings. The infiltrate spreads out from the small, discolored flecks in the annual rings.

The roots of the diseased trees, like the trunk, show the discolored rings, sometimes also the discolored center; moreover, here and there in the discolored center are often larger, brown spots which are just as dark as the specks in the rings. I seldom saw these larger dark spots in the discolored wood in the trunk.

In most cities, to beautify streets and parks they use *Ulmus momentalis* [sic—probably a typographical error for *U. monumentalis*, syn. *U. carpinifolia* 'Sarniensis'], grafted high or low on trunks [understocks] of *Ulmus campestris latifolia* [syn. *U. hollandica* 'Belgica']. The former species seems more susceptible than the latter species. Where the non-grafted *Ulmus campestris*

---

[1]An *infiltration* is a seepage or gradual perfusion of a tissue by a liquid, to wit, the *infiltrate*.

PLATE 1

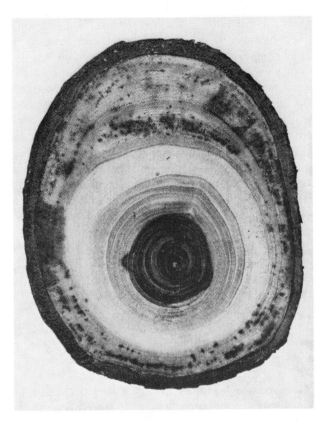

Fig. 1. Cross section [diagonal] of the trunk of a diseased elm tree.
Photo B. Smit. [Note dark dots in at least four annual rings.]

Fig. 2.   Fig. 3.

Fig. 2. Dorsal side, Cattleya beetle. Fig. 3. Ventral side [actually, a lateral view], Cattleya beetle
(enlarged approx. 12 X).

13

latifolia stands, this turns out indeed to show the same disease symptoms, but the number of affected trees is less striking. In large part this results from the fact that little use is made of this species in the cities.

On cutting into the trunk of a diseased *Ulmus monumentalis*, one sees the outermost, discolored ring close to the periphery, close to the cambial layer; with *Ulmus campestris latifolia* this first discolored ring is somewhat deep within. I saw a case in which *monumentalis* was grafted high on a root-stock of *campestris* and in which this *campestris* stem also showed the discolored ring close to the cambial layer, just as though the disease of *monumentalis* had crossed through the graft union into the *campestris*.

By microscopic investigation of various samples of the brown spots, I have found no mold [fungus] mycelial tissue. The walls of the wood vessels there are brown colored, likewise those of the wood parenchyma cells and the ray cells, of both of which last [cell types] the content is entirely brown. The parenchyma cells of the bark are also quite often brown; in that case the cell content evidently is dead and shrivelled to a brown lump.

Insects. From various cities elm bark beetles were sent to us. Mostly we could identify these as the large elm bark beetle, *Eccoptogaster scolytus* F. [later called *Scolytus scolytus*].

In the cases in which I determined the presence of the beetles myself, the trees showed the described syndrome, except in Oud-Beierland; I shall come back to this later.

[p. 6] It is a known fact that such bark beetle infestations are mostly secondary, and so I presume that we are dealing with a weakening of the trees by the disease discussed here or by another disease, before the beetles began their attack.

Moreover, I saw on the leaves of the diseased trees many Typhlociba's (cicadas) and on the branches, especially in the branch forks, red spots formed by the accumulations of mites and mite eggs. The presence of these organisms is nothing special.

Cultures. From material from diseased elms, originating from various places: Hoeven, Knijpe, Oud-Beierland, Tilburg, Delft, Ginniken, Oudenbosch, Tiel, Rozendaal, Schiedam, Rotterdam, Wijk bij Duurstede, Venlo, Opheusden (we received submissions from still other places; however it was impossible for us to make cultures of all), I cultured a number of molds, of which the principal ones in most cases were: *Fusarium* sp., *Phoma* sp., *Botrytis* sp., a mold belonging to the *Cephalosporieae* and one to the *Stilbaceae*.

In one case (Oudenbosch), moreover, a species of *Pestalozzia* showed up, in two cases (Tilburg and Knijpe) a *Verticillium* sp., and in a few other cases a *Didymochaeta* sp.

The fungus belonging to the *Cephalosporieae* we determined as *Cephalosporium acremonium* Corda, although the description thereof does not wholly jibe; the fungus belonging to the *Stilbaceae* looked to us like *Graphium*, perhaps *Graphium penicillioides* Corda.

Since *Cephalosporium* and *Graphium* mostly occur together and are very

14

easy to to get into pure culture on the very first trial, from the inner wood of thick twigs, one of these two fungi, or perhaps both, could be considered the cause of the disease. *Cephalosporium* I have occasionally encountered without *Graphium*; this latter fungus, however, I never found without *Cephalosporium*. Naturally we are doing infection experiments with both of these latter fungi, and also with several others of the molds cultured out of the elms. In view of the little success that has been obtained hitherto with such experiments on woody plants, I haven't much confidence in the success of this.

I want to observe here, moreover, that, even though I myself have repeatedly seen the above-named fungi show up, yet I am not convinced that I am dealing here with a fungal disease. What then the [cause of] disease really is, I don't know; perhaps it is hidden [p. 7] in the soil, or [perhaps] influences of inorganic nature (the extraordinarily severe frost of 1917? the prolonged, dry spring of 1918?) have played a part herein. In all such cases we are certainly on very difficult terrain.

Nurseries. It is important to know whether the disease occurs in the nurseries, because it then could be transferred from there into the cities with the trees.

Mr. ONRUST at Oudenbosch told me that in his vicinity diseased individuals are found among the one-year-old layered plants [a form of rooted cuttings] in the nurseries. According to a writing by Mr. SCHENK, Technical Officer 1st Class with the Phytopathological Service, the elm layers at Knijpe, correctly severed in the fall from the stools (mother plants), in the middle or at the end of March, would show by discoloration of the bark, that something was not as it should be. For many years past, on the average, 20% of the plants have been lost that way. (I cannot say with certainty whether there we are dealing with the disease I described.)

In the different nurseries at Opheusden, where I have seen thousands of young trees, I could find only two small diseased trees, in one man's nursery. These little trees were about eight years old. Nowhere did I see diseased one-year-old layers. Now, in January 1921, I hear from our inspector at Wageningen, Mr. B. SMIT, that in other nurseries at Opheusden, diseased individuals are indeed to be found at this time. I was at Opheusden in the month of November. All trees were bare, and the two small diseased trees could be identified immediately as diseased by their curved twigs. The branches showed internally the discolored, brown rings and I again cultured from them the molds that occurred also with other diseased elms, among others, the *Cephalosporium* and *Graphium*, too. From other, evidently healthy, little trees of the nurseries at Opheusden, I have not been able to culture any molds.

I hope to set up a further investigation this summer, not only at Opheusden but also at nurseries in other places.

Opinions from industry. Through my visits to the places named, as well as from the letters sent to us, I got many data, which I append here:

As soon as the withering begins to arise, the process has, they say, a relatively fast course of events. In a short time the top of the tree is completely withered.

The disease then goes [p. 8] apparently no farther, so that the rest of the tree remains green.

In general they judged that they were dealing with a fungal disease. The infection was thought to come from outside, since they observed the disease only in the top and at the end of the higher branches. This last assumption, however, does not rest on sufficient grounds: because if the roots of a tree are diseased, or if a disease spot occurs in the trunk of the tree, as a result of which the upward water transport is strangled, this expresses itself also by shrivelling symptoms and, indeed, at first by the withering of the extreme tops.

Further, they sometimes saw a tree which at first had been affected, showed nice, freshly developed foliage later in the year, by which the opinion [idea] arose that the tree would indeed be all right again. On investigation I found that with such trees the discolored rings were still present in the wood, so that the disease certainly had not gone.

They did not always agree about pruning. In the one place I heard that pruning was good, the tree thereafter looked more healthy. In another place they were against pruning. In my opinion the pruned trees that I saw were still seriously diseased, as the discolored rings in the wood gave evidence; the severely misformed trees, which had become much smaller in extent, could perhaps still live for some time on their food reserves; moreover, the water carried up was probably enough for the now diminished crown. I was told that unpruned *Ulmus campestris latifolia* did not show the disease. In one of the places I visited I saw such a row, which apparently were healthy trees. I did not have branches taken off these, so I cannot say whether the wood discoloration was present.

In some cases trees in newly laid-out streets, thus planted in freshly dug-up earth, seem not to be diseased, while trees sold at the same time, placed in old soil (city nurseries; earlier laid-out streets), do seem to become diseased. In such a case one would be tempted to think of soil questions.

In one place, they think that the cause must be sought in the raising of the streets [i.e., fill]. The roots there are often more than 1 meter deep in the ground, which naturally is not advantageous for the health of the trees.

Also, occasionally the idea was proposed that the gas from the gas pipes in the ground would be the cause of the disease. [p. 9] With this, however, typical symptoms occur (loosening of the bark, turning blue of the roots), which were not present in the cases investigated by me. One respondent thought that where brown coal [soft coal] so often was used as fuel in recent years, the gasses and vapors that developed from this could have influenced the trees.

In a few cases they associated the disease with the strong bloom of elms in the spring.

Control. So long as nothing more definite is known to us about the disease, we cannot speak of a control, either. Currently our recommendations have been to cut away the dead tops and [dead] branches as far as possible, and to paint the wounds with tar or carbolineum, against infection, not to prune further, and to await what will happen to the elms this summer.

Where boring by the elm bark beetles is to be seen and this attack still is not so heavy that one need fear for the condition of the tree, painting the trunk and the large branches with a brush with carbolineum (about 30%) is advisable, about May, as the first beetles fly. People think they have observed that the beetles die if they land on the greased trunks. I don't know whether this is right, but in any case the female beetle seems to be prevented from laying her eggs.

Oud-Beierland and Venlo. In Venlo Mr. SCHROEVERS noticed that the roots of many, mostly very old, trees were totally rotten. These trees were, in general, in bad shape, yet he did not get the impression of dead tops, as in other communities.

In Beierland I saw a row of very old trees, two of which were already dead. The bark was loose on all these trees; from the dead trees I could pull off the bark from the wood, from below to above, as a loose coat. Such a thing I have observed in none of the other places. Hundreds of bark beetles had dug their tunnels under the bark, and were a threat to the trees standing in the vicinity, in which I found bark beetle holes everywhere, already. The disease of this row of trees reminded me of gas poisoning. Since I have not seen the roots, and it was not possible at that moment to cut a branch of the tree in order to see whether there was discoloration of the wood, I would not dare to give further judgement about this case.

Along the water in front of the Technical [Vocational] School at Oud-Beierland [p. 10] a number of older trees were also badly affected by bark beetles. These trees did show somewhat more the disease picture that occurs elsewhere: in the wood and in the roots I saw the well-known colored rings, also the curved, withered twigs from sprouted, dormant buds on the trunk. However, I did not see distinct withered tops in the trees.

Conclusion. From the above it turns out that we do not yet know much about the disease. We now do know the syndrome. We know further that the disease probably first could be seen in 1919. The place of the discolored annual rings in the wood, however, shows that the beginning of the disease must have arisen earlier. In most cases one encounters discoloration in the annual rings of 1920, 1919, 1918. See figure 1. (The dark center in the trunk has nothing to do with this disease.) Occasionally we have some presumption that the annual ring of 1917 was affected also.

This year we shall examine accurately the entire disease process yet again, and await the results of the infection experiments. If these do not succeed, then we shall have to repeat them annually under other conditions, on different dates, etc., which can become a very prolonged investigation. The investigation can be just as long if it turns out that we must seek the cause underground. The elms that are now diseased do not benefit from that research at the moment, and if the disease does not disappear again in the same inexplicable way in which it showed up, I fear that the trees will go on downhill farther, and finally will die.

In my opinion, no result can be expected from [attempts to] control the disease by spraying with some fungicide or other.

At the end of this article, a word of thanks to Mr. SCHROEVERS for his interest and help with this preliminary investigation.

Dina Spierenburg,
Phytopathologist, Phytopathological Service
Wageningen, January 1921.

An Unknown Disease Among the Elms [II]

Spierenburg, Dina

1922. *Een onbekende ziekte in de iepen, II.* Plantenziektenkundige Waarnemingen III. [Also published as:] Verslagen en Mededelingen van den Plantenziektenkundigen Dienst te Wageningen (nr.24): 1–31, 3 tabellen, 1 tekening, plaat I–IV, fig. 1–6, en 1 kaart.
[Plant Disease Observation III. Also: Reports and Communications of the Plant Disease Service at Wageningen (number 24): pages 1–31, 3 tables, 1 drawing, plates I–IV, figs. 1–6, and 1 map.]*

[p. 3] As sequel to the article that appeared in January 1921 in Communication of the Plant Disease (then still Phytopathological) Service, PLANT DISEASE OBSERVATIONS I, entitled "An Unknown Disease Among the Elms," a review will be given below of what has become known during the course of the year 1921 about this peculiar disease in the elms.

Distribution. Just as in 1920, in 1921 reports were sent to us from very different parts of the country, or questions were posed about the elm disease occurring in those reaches of the country. I have laid out a map of the places in which (or in near vicinity of which) we know for sure that the elm disease was observed (Plate I),[1] on which is not indicated if [whether or not?] the diseased elms occur on connecting roads between two towns.

In general we can say that in the province of North Brabant the disease occurs wherever elms grow. Further we see a severe outbreak of the disease in the Betuwe, and in different places in the province of South Holland; the disease reigns most seriously in Rotterdam and vicinity. In none of the other places in the country have I observed a heavier outbreak than in Rotterdam.

In North Holland the disease occurs as far as Alkmaar. In the course of the summer a few diseased trees were observed in Amsterdam. I am curious as to whether the disease is in an initial stage here, and whether it will assume a more serious character the following year.

In the Northeastern provinces of our country the disease occurs sporadically; in the late summer an initial occurrence of the disease was reported from several places; in Friesland, by a personal visit, I could establish the presence of the disease only at a nursery in Joure.

[p.4] It seems, in a sense, as though the disease occurs in a more severe degree in the South than in the North. In this connection, however, it must

---

*[Translated by F. W. Holmes, 3/1985. The roman numeral II in the title is in brackets because it was left out of the original title, appearing only on the paper wrapper. That caused many people to think they had already read the article and therefore to overlook the remarkable things that were reported here.]

[1]The writer solicits the favor of reports about the occurrence of the elm disease in places not marked on the map.

not be forgotten that in the South a whole lot more elm trees are cultivated and planted than in the North.

Abroad. In order to find out whether **the** elm disease is also known *abroad*, we have contacted different foreign phytopathologists, forestry specialists, heads of Plant Disease Services, etc.

The result of this correspondence is that the disease seems to reign throughout *Belgium*: "... and I hope to be, in the course of this October[1], in a position

[1]October 1921.

PLATE I

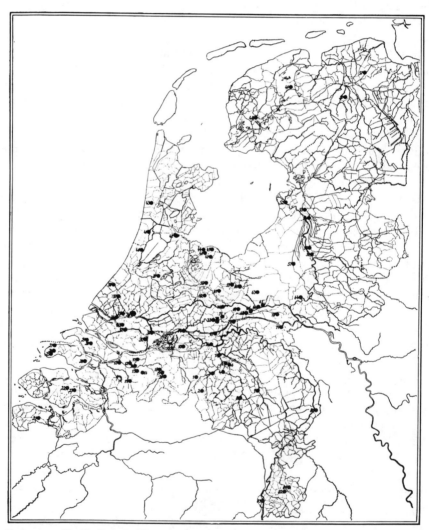

Teekening : W. H. Ruisch, Wageningen.

to supply information to you on this subject," they wrote. To date we have received no further information.

In *France* the disease occurs only to the North of the Seine River: "In France the elm disease was reported to me in 1919 and succeeding years in Champagne and in the Argonne. Since that time its presence was pointed out to me in several regions, all of which are located to the north of the Seine valley," Mr. FOEX, Director of the Plant Pathology Station at Paris, writes to us.

In *Denmark* and *Sweden*, according to the phytopathologists with whom we corresponded, and a few of whom visited our country so that we could discuss the disease with them, **the** elm disease does *not* occur.

Official reports from *Germany* state that they do not recognize the disease there. However, Directors of municipal plantings from The Netherlands who have traveled along the Rhine, and also others who had experience in the field of **the** elm disease, *think* that the disease does occur in Germany.

Places Shown on the Map

39. Aarlanderveen	74. Gorkum	15. Oudenbosch
76. Ammerzoden	30. 's-Gravendeel	32. Oud-Beijerland
42. Amsterdam	38. 's-Gravenhage	
57. Apeldoorn		50. Peize
	26. Haamstede	33. Poortugaal
49. Bedum	3. Haaren	
58. De Bildt	41. Haarlem	27. Renesse
4. Boxtel	61. Hardenbroek	65. Renkum
19. Breda	9. Helmond	36. Rotterdam
71. Buren	7. 's-Hertogenbosch	17. Rozendaal
25. Burg	40. Hillegom	
46. Bussum	2. Hilvarenbeek	60. Scherpenzeel
	82. Hoensbroek	35. Schiedam
43. Castricum	16. Hoeven	5. St.-Michielsgestel
	62. Houten	48. Sneek
73. Deil	45. Huizen	81. Staatsm. "Emma"
37. Delft		14. Standdaarbuiten
56. Deventer	69. Ingen	
55. Diepenveen		51. Ten Boer
64. Dieren	52. Kampen	77. Tiel
13. Dinteloord	23. Kapelle	1. Tilburg
28. Dirksland	68. Kesteren	
31. Dordrecht	11. Klundert	80. Venlo
10. Dussen	18. Kruisland	34. Vlaardingen
63. Ede	47. Laren	66. Wageningen
8. Eindhoven		54. Windesheim
78. Elst (O.-B.)	44. Naarden	59. Woudenberg
	79. Nijmegen	70. Wijk bij Duurstede
12. Fijnaart		
	83. Maastricht	21. IJzendijke
72. Geldermalsen	29. Melissant	
6. Gemonde		75. Zalt-Bommel
20. Ginniken	24. Oosterland	53. Zwolle
22. Goes	67. Opheusden	

In *England* according to what they wrote to us, the disease seems not to prevail, but then again the elm planting there is very far from general. One of our phytopathologists, Mr. SCHROEVERS, who twice spent a number of days in England in 1921, has not been able to confirm the disease there.

So long as no official publications appear abroad about the disease, we could only convince ourselves by personal investigation whether the disease occurs in other countries. *None of our own officials has seen the disease in one of the surrounding countries.*

Species of elms. Various elm species can be affected by **the** disease. Now, about the concept "*species*" a whole lot should be said [p. 5] which I cannot go into further here, and which I prefer to leave to others, but we have been able to confirm **the** disease in all *diseased* individuals that were sent to us under the name of a *particular species.* Of the "*species*," we want to name here: *Ulmus campestris latifolia* [syn. *U. hollandica* 'Belgica'], *Ulmus monumentalis* [syn. *U. carpinifolia* 'Sarniensis''], *Ulmus campestris suberosa* (cork elm), *Ulmus americana* (American elm), *Ulmus campestris aurea* [syn. *U. carpinifolia* 'Wredei'] (gold-yellow-leaved elm). To date I still have not found the disease in *Ulmus vegeta* [syn. *U. hollandica* 'Vegeta']. (However, according to Mr. LEONARD A. SPRINGER, horticultural architect in Haarlem, this variety is not very suitable for city plantings.)

Disease appearance. External. In the past year it struck us that various disease appearances occur, which can be fitted into three groups which cannot be completely separated:

1ˢᵗ. The appearance described on page 4, "Plant Disease Observation I," thus: *withered top*, leaves crisp [brittle], commonly small, *recurved twigs* (see Plate II, fig. 2, and Plate IV). In a serious case the desiccation of the leaves moves downward in a very rapid tempo, so that in the middle of the summer the diseased tree is utterly dried out. With very hot weather this process can complete itself in one day. In most cases the leaves of the lower branches of the tree stay green for the whole summer, against which the withered leaves in the top [are in] distinct contrast.

The situation described here was already known in 1919 and 1920. The trees with this appearance all (?) *die.* Of the trees affected in this manner in 1920, I have myself seen not a single one that still lived at the end of 1921. A report of the recovery of a tree was sent to us.

Most of the trees with this appearance are *not older* than 30 years.

2ⁿᵈ. An appearance that occurs with huge, *old* trees. The trees at first look healthy with (almost) normal foliage. Suddenly there shows here and there a branch with withered foliage, which is followed by more. Only *later does the top dry out.* In some cases this process, too, goes faster and such a big tree dies within a few weeks, without anyone having noticed anything of the disease earlier in that tree.

3ʳᵈ. The chronic disease picture, of which again old trees are the principal victims: the patient has looked bad for a longer time. In most cases we hear

PLATE II

Fig. 1.

Micro-foto en foto B. Smit, Wageningen.

Fig. 2.

that such a tree [p. 6] already has done poorly for several years. The trees leaf out with difficulty in the spring, the leaves remain small. After "St.Jan" [the feast of St. John the Baptist, celebrated on June 24; in folklore this defines midsummer but at present June 21, the summer solistice, is considered the beginning of summer] the tree begins to look as though it is autumn. The inner foliage falls and the remaining leaves are dry by mid-July and fall off early. The tree has had sparse foliage through the whole summer.

PLATE III

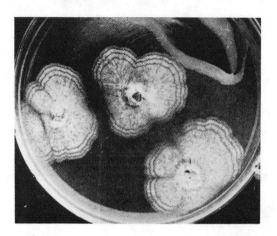

Foto B. Smit   Fig. 3.

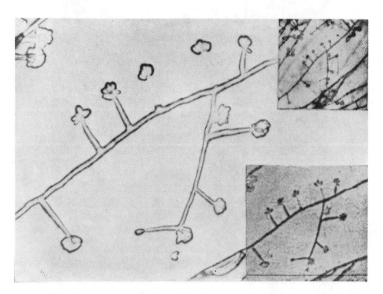

Micro-foto B. Smit.   Fig. 4.

24

We still do not know satisfactorily how it will turn out for the trees of the 2nd and 3rd type of case.

Disease appearance. Internal. In "Plant Disease Observations I," on page 4, the rings of discolored stippling in this wood were mentioned: the thicker branches, the trunk and the large roots show one or more such rings of *discolored stipples* [dots]. In seriously affected cases, it is no longer dots but entire bands of discolored wood: this especially with older trees, whose wood sometimes seems to be more diseased than the outward aspect of the tree would suggest (*2nd* and *3rd* disease picture).

The rings of dots are seen in the annual rings 1921, 1920, 1919, 1918.[1] The discoloration in the 1918 annual ring was observed only one or two times last year; in the 1917 annual ring, no additional cases were seen. One mostly sees the *discolorations* in the *last three annual rings*. In the trunk, where the rings of dots are closest to each other, the color of the dots is less striking, since the wood of elm trunks ordinarily is somewhat darker than that of elm branches and elm roots. The outermost annual rings in the trunk often lie so close to each other that counting the rings becomes difficult.

[1]Once, indeed, we did wonder whether [or not] the discoloration didn't occur still deeper than in the annual ring [of] 1918, 1917. In the cemetery at Rotterdam, we were present at the felling of a tree about 80 years old, which clearly showed a dark dot in some earlier annual rings (1913), which again made us think of the disease. Later a thick trunk slice of this tree was sent to us. When shaving off this slice we found additional discolored spots, about 2 cm. long and 1 to 3 mm. broad. For this, see Cultures, page 19.
[The heading "Cultures" begins on her p. 13, but cultures from this tree are reported at the bottom of her p. 19, where it appears that *C. ulmi* (her "*Cephalosporium-Graphium*") grew from *some* cultures made from rings of 1912-1915. Alas, the author did not say from which rings, so this confirmed infection may have occurred as late as 1915. But note what she says on her p. 16 about isolating the DED pathogen from rings of 1912 and 1913 of elms in the town of Renkum! Clearly DED arose before World War I.]

PLATE IV

Foto: A. Lebbink, Rotterdam.   Fig. 5   *Foto: Th. J. de Vin, Elst.*   Fig. 6.

The roots, if they are *not* internally rotten, which often occurs with elms, look entirely sound, bright yellow-white, with discolored dots in the last annual rings.

General remarks about the occurrence of the disease and about the disease picture. Taken in general, the disease in 1921 has been just as serious, perhaps even more serious than in 1920. [p. 7] That the disease *seemed* to be more serious in 1921 in some places, in my opinion arises from the fact that people did not notice the *ailing* trees in 1920, when they did not yet pay as much attention to the diseased elm trees. In 1921 these trees *seemed* suddenly to have become seriously ill, while really they already *had been suffering* in the preceding year. In the trees that *declined* in 1921, when people once realized that there prevailed a disease among the elms, they immediately investigated the branches and then found the brown rings, thus the disease. They now know that the trees are diseased, whereas earlier the trees *were diseased*, without their knowing it.

We don't yet have experience about the course of the disease in declining trees. Most of those practicing [arborists] don't believe that the trees will stay alive.

Where the disease first shows its presence by the withered top, the process mostly procceds as usual, and quickly leads to death. It is notable that the withered leaves always remain *attached* to the branches for a *very long* time.

It often happens that buds remain dormant, especially in older trees. Sometimes such buds do leaf out later in the year, but the twig with leaves that arises from this mostly withers speedily. The conducting tissues apparently lack the ability to keep providing the young wood with the necessary sap. In very many cases one finds shoots pushing out right on the trunk, so-called water-shoots [epicormics], a generally occurring symptom on other diseased trees as well. The water-shoots formed here wither quickly after the sprouting.

Trees that wither early in the spring do leaf out again later, often with fresher and larger foliage than at first, but these newly formed sprouts wither also. If the leaves of a tree shrivel later in the year, then resprouting is rarely seen.

They wrote that on *young* trees in the spring sometimes *discoloration of the bark at the buds* was visible. I have seen that myself, too, but the fact is too unusual to be able to associate it with **the** disease. I have also observed it with an *older* weeping elm. Later in the summer one can no longer see a difference between the originally darker bark under the buds and the rest of the bark.

Very old trees are less susceptible than young or half-grown trees. Perhaps, too, that only *seems* so and the trees are indeed just as susceptible; still, the process proceeds slower in such trees, possibly in connection with the accumulated reserve food and greater surface of water transport.

[p. 8] In general, on strongly affected trees one still finds the buds for the following year, but these are small, sometimes already dried out. With less strongly affected trees one does find thicker buds, yet these then turn out to be blossom buds, an attempt by the diseased tree to provide for the survival of the species, if possible, by blossom- and seed-production.

If one has just cut off the branches of a tree, or if one cuts up a tree, the inner wood is very moist, but quickly dries. The dark specks [dots] in the [moist] wood become less distinct in the dry wood or disappear, inasmuch as elm wood, exposed to the open air, always discolors darkly to some degree, and the dots then no longer contrast well with the healthy wood. A ring of **discolored dots,** which is **clearly visible** upon the cutting up of a tree, **never** wholly vanishes. Where the affliction is such that there are many dark dots and rings present in the wood, these continue to keep the brown color even after preservation of the wood in the laboratory for months on end.

Sometimes a brown ring occurs in the branches, exactly between bark and wood. Such branches speedily die. Mostly, however, a very thin layer of wood can be found between the bark and the last ring of dots, which are perceptible here and there through the wood and also occasionally border directly on the bark. In this last case the formation of summer wood on those spots presumably is not yet completed. Such branches then stayed alive the whole summer, and bore green foliage until late in the year. **Outside the last ring of dots,** then, **healthy wood** can still occur.

Sometimes the ring of discolored wood does not run through the whole branch. In a few places, upon cross section, the wood is normal in color, without dots. Also lower in the trunks and in the roots the discoloration is sometimes absent. It also [has] happened that the discoloration is not present halfway up the trunk, but shows up again lower, at the roots. This, however, is sure: that with a *young,* diseased tree, **discoloration in the branches in the top** can always be found, and that with **thoroughly diseased,** *older* trees **discoloration of the wood is present everywhere,** even in the thinnest twiglets and in the thinnest roots.

Infectivity. In many cases our judgement is asked about whether the disease is or is not infectious. About this we can say the following: a few times the location of the diseased trees suggests to us that the disease originated from a focus of infection. On the other hand, however, we often see [p. 9] some healthy (?) trees standing in an otherwise wholly diseased row. Also we know groups of trees, one or more individuals of which are affected while the others seem healthy to date. It will have to be seen later whether the trees now apparently healthy nevertheless in a later year are going to show the disease picture.

There is **no certainty** about the **infection.**

Insects and other animals. Except in those cases in which the elm disease was still in an initial stage, the elm bark beetle occurred rather generally this summer, mostly the large elm bark beetle, *Eccoptogaster scolytus* F. In this winter, 1921–1922, many diseased trees, which this summer we imagined still beetle-free, turned out to be affected in strong degree by the bark beetles. We now find under the bark, besides the beetle larvae, fly larvae as well. These are longer, thinner and much more wriggly than the beetle larvae. What fly species we are dealing with here will appear later, when the larvae will be cultured to imagos.

Typhlocybas were encountered in great numbers, primarily in the late summer, just like last year; moreover, [we] very generally [encountered] mites (*Oribatas*), too, and mite eggs, both red.

Further, on the spot where the 1921 bud broke open, thus at the boundary of the sprouts of 1920–1921, in the grooves which are always present at such a sprouting bud, I found very tiny, red bodies, mites of the group *Eryophyidae*. Dr. A. C. OUDEMANS at Arnhem, who was so kind as to establish the correct name for us, determined them as *Anthocoptes* sp., a species with 12 dorsal semicircles [crescents], related to *Anthocoptes loricatus* NALEPA with 8 and *Anthocoptes salicis* NALEPA with 10–15 dorsal semicircles.

During the summer they were present in larger number than in the fall and in the beginning of the winter, but they still can be found now.

Nurseries. In the nurseries the disease could be found much more than in the year 1920. We already heard, in February and March, that **the** disease could then be clearly observed in various nurseries. The nursery operators take away the diseased trees, or they decapitate them, whereupon, seen superficially, there is little disease present, but if one walks through the rows of trees in the nurseries one clearly sees how, by new interplanting, the rows have become uneven or how, among the healthy **[p. 10]** trees, many trees are standing with sawed-off tops.

Diseased layers (sinkers) I have never seen in the nurseries. Also I have never noticed that one-, two-, or three-year-old trees, which I needed for infection experiments, were diseased. The youngest trees that I have seen diseased were certainly 4 years old in 1921.[1]

Nowhere have I encountered diseased stools (mother plants) [for layering]. A correspondent did write to me that in one of the nurseries present in our country diseased stools are to be found. However, I have not validated this report.

Microscopic investigation. In the microscopic investigation, I found that the discolored rings of the wood always occur in the *wide* vessels, thus in the *spring wood*. Discolored spring wood regularly occurs also with the more inward discoloration (1912 and 1913) [see comment in footnote on p. 25, p. 6 of original]. The trees that show a discolored last annual ring (1921), and are dying, have the discoloration in the spring wood and thereafter mostly have formed new wood, so that, with such trees, the brown 1921 discoloration gleams hazily out through a very thin wood-layer.

As far as the discoloration itself is concerned, all that the microscopic investigation taught [about it] is that, in the discolored places, the walls— of the wood vessels, wood fibers, wood parenchyma cells, and pith-ray cells, whether or not discolored—are very distinctly visible, standing out against

---

[1]From a shipment of little elm trees which we [have] just received (January 1922) for infection experiments, **all** one-year-olds turn out to be afflicted by **the** disease.

a mostly brown wall coating and brown contents. The pith-ray cells mostly contain large starch grains which, if the pith-ray cells chance to lie in a discolored ring, are entirely brown. At transition places one finds pith-ray cells with grains that are still partly good, among the brown mass.

In the walls of all wood elements, many distinct pits occur; everywhere in the vessels one sees tyloses, which in most cases have become brown.

As is described in "Plant Disease Observations I," page 5, the bark cells sometimes have a brown content. Macroscopically, if one pulls off the outermost bark, one finds a small brown speck here and there in the green bark.

Sometimes, in discolored cells, as much in wood as in bark, [p. 11] *bacteria* occur (see Cultures, page 21). A longitudinal section of the wood at a discolored speck reveals macroscopically that the discoloration is not uniform, but that sometimes very dark, almost black longitudinal stripes are visible adjoining lighter brown ones. [It is] primarily in a microscopic section at the darker places [that] one finds the bacteria. In a cross section through the wood, if one happens to meet such a dark speck, it is evident, even macroscopically, that a small, dark droplet squeezes upwards: such a droplet contains a good many bacteria; the lighter brown places contain fewer bacteria.

Practically speaking, we have not been able to show any fungal tissue, notwithstanding many sections made by me, as well as by the other phyto-pathologists of our Service, Messrs. SCHROEVERS and MAARSCHALK.[1]

A few times we did manage to get to see some fungal hyphae, especially with young wood in which little brown discoloration as yet occurs, but we haven't any certainty about the course of that mold hyphae, and the cases [p. 12] are still too sporadic to be able to say with certainty whether we are dealing with the fungus of **the** elm disease.

We applied different staining methods, to try to get to see the mycelium more clearly. The dye method we regularly used—lay the section in cotton blue and, to promote the staining process, parboil the whole thing, whereupon the mycelium (mold thread tissue) always gets colored distinctly blue and

---

[1]Professor A. **TE WECHEL** at *Wageningen* was so kind as to call to my attention the similarity in the microscopic picture between the prevailing elm disease and the (false) heartwood formation of beeches (false in that beech trees actually form no heartwood). In the *Zeitschrift für* Forst- und Jagdwesen 34th volume, 1902, on page 596, there is an article by G. HERRMANN: "Ueber die Kernbildung bei der Rotbuche" [On heartwood formation in the red beech]. HERRMANN, too, could seldom show a fungus in discolored wood of the heart. Plate IV, figs. 8-14, of the article, gives pictures of macroscopic cross sections of the wood, which could serve just as they stand, as microscopic pictures of our diseased elm wood. Further, plate III gives a photographic picture of a beech trunk which shows the same sort of discolored rings as our diseased elm trunks. On page 614 we read, about this: "The forming of characteristic brown rings in the sapwood of the trunk remains unexplained, 11, fig. 4. Anatomically they have wholly the character of false heartwood. The vessels are closed by tyloses and wound gum plugs, in part also filled with gray crystalline masses of calcium oxalate. Globular and granular, partly also homogeneous, protective gum masses occur in the parenchymal and medullary ray [pith ray] cells. The cell walls are colored bright brown. No mycelium could be detected, even after staining..."
"The occurrence of the same [symptoms] in heartwood-free trees proves that these peculiar brown bands have nothing to do with the heartwood formation, as I had opportunity to observe here in Wirthy. What, however, has caused the formation of the same, I could not, as stated, find out."

then is easy to recognize among other plant tissues, a method which normally never fails—gave no success. Other staining methods used by us:

a. Staining with fuchsin. Mycelium becomes red.

b. Staining with ruthenium red. Mycelium becomes strongly red colored.

c. Staining with saffranin and light green. The hyphae (mold threads) stand out distinctly green against the rest of the tissue. (Journal of Agricultural Research, volume XI, number 6, "Diagnosing white-pine blister rust from its mycelium." REGINALD H. COLLEY.)

d. Stain with picric acid and Delafield's hematoxylin. The mycelium gets purple-blue colored; the other tissues are citron-yellow. "Silver Leaf Disease." J. BINTER. Royal Botanic Gardens, Kew, Bulletin of Miscellaneous Information, 1919[6-7].

e. First stain the sections with Delafield's hematoxylin, thereafter with eosin. The mycelial strands then are rose, in contrast to the purple-blue of the other tissues. (See Phytopathology, Volume 1[4] 1911, "The differential staining of intercellular mycelium." ELIAS J. DURAND.)

When, however, we do encounter mycelium, it certainly has been just as visible with cotton blue as, or better than, by staining according to one of the other methods, so that we prefer to keep to the simple and quickly performable cotton-blue method.

We did find that the mycelium of *Cephalosporium* sp., the mold which can always be cultured from discolored, diseased elm wood (see Cultures, page 13, and Plant Disease Observations I, page 6), does not absorb a dye as easily as other mold tissue usually does.

I have tried yet another way to get to know whether and where the mycelium occurs in the wood. For that, I took very thin sections and put these into a hanging drop of water under the cover-glass of a *van Tieghem* cell.[1] Then, within one or two days the **[p. 13]** *Cephalosporium* mycelium came in sight and grew from the thin slice of wood, further out along the cover-glass. We *never* succeeded in discovering exactly where the mycelium came from.

Cultures. Molds. Many more cultures were made in 1921 than in 1920. Since the disease picture has gradually become very familiar to us, and since we had found that in the cases where we dealt with discolored wood of *well-diseased* trees, we always could culture out the fungi *Cephalosporium-Graphium*, we applied ourselves in 1921 primarily to the plating out of cultures from wood of trees about which we doubted whether or not **the** disease was involved, [and] further from wood of elm species in which **the** elm disease was still unknown to us (for those species see page 4). [The Latin names, however, appear on her page 5.]

The principal part of our investigations in 1921 involved a street with diseased trees in Renkum, research on which was permitted by the Messrs. BEUKER

---

[1] A *van Tieghem cell* is the name given to a small, closed-off space which one gets by fastening onto a microscope slide a little glass ring a few mm. high, and closing the whole thing off by a cover-glass. Instead of the glass slide with little glass ring, a glass slide can also serve in which a hollow cavity is ground, over which one lays the cover-glass.

in Renkum, directors of the paper factory *"United Royal Paper Factories of the Firm Van Gelder Sons,"* and owners of the trees that stand on the road to the factory. For their kindness I give Messrs. BEUKER, with pleasure, my thanks. The avenue (see Plate II, fig. 2) consisted of 21 trees on one side, 19 trees on the other side, according to the drawing below.

```
                                     West
                            South---+---North
                                     East
               Porter's dwelling
 1   2   3   4   5   6   7   8   9  10  11  12  13  14  15  16  17  18  19  20  21
 o   o   o   o   o   o   o   o   o   o   o   o   o   o   o   o   o   o   o   o   o

           Factory fence <---------              ---------> State Highway

    22  23  24       25  26  27  28  29  30  31  32  33  34  35  36  37  38  39  40
     o   o   o        o   o   o   o   o   o   o   o   o   o   o   o   o   o   o   o
```

The trees are all about 30 years old and planted 12 years ago. In the table printed on the following pages [Table 1] I have summed up what was observed on the trees.

[p. 16] We could determine the discolorations encountered in the wood in the directions North, East, South, and West by boring in the wood of the trees from those directions with an increment borer (a drill of PRESZLER).

For completeness sake, I have included these borings in this report. We carry out wood borings in order to be able to determine, without cutting down or damaging the entire tree, whether a particular tree shows internal discoloration. The drill makes an opening of only about 3 mm. diameter. If one by chance finds no discoloration at a drilled location, and if the tree nonetheless has the appearance of being affected by **the** elm disease, then one usually does encounter discoloration when boring somewhat higher or lower.

In the street in Renkum 4 holes were always drilled in each tree, precisely in the directions North, East, South or West, and always at 1.25 meters above the ground.

These borings went so deep into the tree that I could observe at the same time whether in the innards of the tree there was a darker heart present, which is reported in the 7th column. The year 1912 or 1913 indicates that in the discolored heart in the year 1912 or 1913 a distinct darker stripe of discolored summerwood is present.[1] The rest of the internal discoloration is uniform.

One of the trees (number 2) I have gradually cut down with the help of my colleague, Mr. H. MAARSCHALK. We went regularly to Renkum, cut

---

[1] Cultures of this wood sometimes gave *Cephalosporium-Graphium* [sic].

TABLE 1
REVIEW OF THE TREES IN THE STREET IN FRONT
OF THE PAPER-FACTORY AT RENKUM

Tree number	Summer Observations			Winter Observations						
	Appearance in summer 1921	Wind direction toward which the disease mainly spread, summer 1921	Elms sprouted from the trunk	Trees affected by bark beetles	Curling of the twigs	Internal discoloration	Discoloration in the most recent annual rings			
							N	E	S	W
1	G	—	—	—	—	+1912	.	:	.	—
2	(1)	—	—	—	—	—	—	—	—	—
3	2/3(2)	—	—	:	:	:	+	+	:	+
4	G	—	—	:	:	:	:	.	:	—
5	+	N	—	+b	+	+1912	—	+	—	—
6	:	NW	—	:	—	+	—	.	—	—
7	+	—	—	+b	+	+1913	—	.	—	+
8	2/3	—	—	—	t:	+1913	:	+	+	+
9	.	NE	—	—	.	—	—	.	—	.
10	G	—	—	—	.	┤1912	—	.	—	.
11	2/3	—	—	—	t:	—	:	+	:	.
12	+	NW	—	+	+	+	—	+	.	+
13	+	NW	—	+	+	+	—	.	—	.
14	1/3	SW and N	—	—	+	:	—	—	—	:
15	+	—	—	—	+	+	.	.	:	.
16	:	NW	—	—	.	+	—	.	—	—
17	+	NE	—	:	.	+	.	.	.	.
18	:	—	"	+b	+	—	+	+	:	+
19	+	NE	"	—	+	+	+	+	+	+
20	+	SW	"	+	+	+	+	+	.	+
21	1/2	SW	"	—	.	+	.	.	.	.
22	+	SW	—	—	+	+	+	+	.	+
23	2/3	—	—	—	+	+	+	+	.	+
24	G	S	—	—	.	+	.	:	—	:
25	1/2	—	—	—	.	+	.	+	.	+
26	G	—	—	—	.	+	—	—	—	—
[p. 15]										
27	+	—	—	:	t:	+	.	:	.	:
28	G	—	—	—	—	+	—	—	—	—
29	+	S	—	—	t:	+	+	+	.	+
30	+	—	—	+b	t:	+	+	—	.	+
31	+	NW and S	—	+b	+	—	+	+	.	—
32	+	—	—	+b	+	—	.	:	.	:
33	1/3	—	—	—	+	+	+	+	+	+
34	1/3	—	"	—	:	+	+	+	+	+
35	1/3	—	"	+	+	+	.	+	+	.
36	1/2	S	"	—	+	+	+	:	+	:
37	+	—	—	+	+	—	—	:	.	.
38	:	—	"	—	t:	—	+	+	+	+
39	G	—	"	—	:	+	—	.	—	—
40	+	—	"	—	+	—	+	.	+	+

[continued on next page]

TABLE 1 continued

Tree number	Summer Observations		Winter Observations							
	Appearance in summer 1921	Wind direction toward which the disease mainly spread, summer 1921	Elms sprouted from the trunk	Trees affected by bark beetles	Curling of the twigs	Internal discoloration	Discoloration in the most recent annual rings			
							N	E	S	W
TOTALS:										
	+ 16	N, NW,	+ 0	+ 11	+ 18	+ 28	+ 14	+ 17	+ 8	+ 16
	: 4	or NE	" 10	: 5	: 3	: 3	: 3	: 6	: 5	: 5
	. 1	= 10	. 0	. 0	. 8	. 0	. 9	. 12	. 14	. 7
			− 29	− 23	− 4	− 8	− 13	− 4	− 12	− 11
	2/3 = 4	S or SW			t: 6					
	1/2 = 3	= 8								
	1/3 = 4									
	G = 7	All wind directions = 23								
	39		39	39	39	39	39	39	39	39

(1) Cut down, see page 16 and table 2, page 18 [pages of original text]
(2) The fractions indicate what portion of the tree was diseased.
G = healthy [the Dutch word is *gezond*]
+ = affected severely (discolored, etc.)
: = affected moderately (discolored, etc.)
. = affected very slightly (discolored, etc.)
b = bark has nearly entirely disappeared
t = top
" = sprouted on trunk

off the pieces of wood which we considered we needed, studied them, and took them back to the laboratory in order to make the cultures there. Thereby we had the advantage that the trunk and branches remained alive and that we had fresh material continually at our disposal. A review of the cultures from the wood and from the bark of tree 2 is given in the plate below [originally on p. 17], which provides a schematic presentation of the tree, and in table 2, page 18. The tree was 9.75 meters tall; the branches began 2.31 meters above the ground.

When we began with the study of the tree, about half of the crown was entirely withered and, lower down, withering but in large part still green branches were visible. Where we cultured out the molds *Cephalosporium-Graphium*, that was always from **discolored** wood of the last annual rings (1921, 1920, 1919). In most cases we could find only one or two annual rings with discolored dots.

[p. 18] If the wood was **not discolored,** then there was ordinarily **no** mold culturable from it, and certainly **never** *Cephalosporium-Graphium*.

From the **bark** cultures I also **never** got *Cephalosporium-Graphium*. Bacteria

Schematic Presentation of Tree 2
of the Lane in Front of the Paper Factory at Renkum

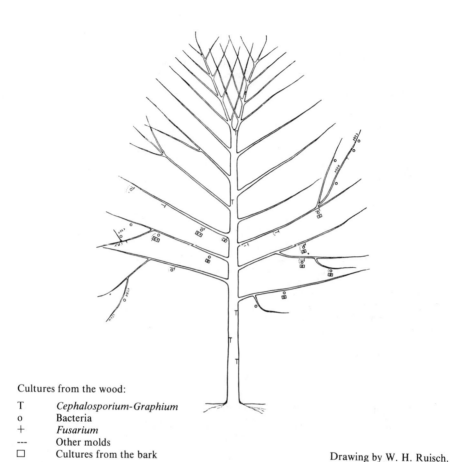

Cultures from the wood:

T   *Cephalosporium-Graphium*
o   Bacteria
+   *Fusarium*
---   Other molds
□   Cultures from the bark

Drawing by W. H. Ruisch.

[p. 18 continued] came from the wood as well as from the bark.

A portion of the diseased wood of the tree is under chemical investigation. This research is not yet completed.

The discoloration in the wood became continually less from the crown downwards in the trunk, and utterly stopped about 1 meter above the ground. At 2 1/2 meters to 1 meter above the ground the wood discoloration actually could still be followed only on one side of the tree (Northeast), as a narrow discolored line. [p. 19] From the basal piece of the trunk, about 5 cm above the ground, we plated out cultures of the wood of the different annual rings. Discoloration could not be found in the wood of the basal piece of the trunk.

TABLE 2
TREE 2. LANE IN FRONT OF PAPER FACTORY AT RENKUM. FUNGI 1921.

	Number of cultures	Cephalo-sporium	Graphium	Fusarium	Cyto-sporina	Monilia	Dematium	Phoma	Ramu-laria	Bacteria	No mold or bacteria
11 culture series[1]		7[2]	6	4	1	2	4	1	1	11	5
Discolored wood	12	12[3]	10	2	1	1	—	—	—	10	—
Nondiscolored wood	9	—	—	1	—	—	1	—	—	7	3
Limited/doubtful discoloration	6	2	1	—	—	1	—	1	—	4	—
Twigs of 1920	8	1	—	1	—	—	1	—	—	5	2
Twigs of 1921	3	—	—	—	—	—	—	—	—	3	—
Leaf petiole (1921)	2	—	—	—	—	—	—	—	1	1	1
Bark	9	—	—	3	—	—	3	—	—	6	1
Interface of bark and wood	3	—	—	1	—	—	1	—	—	1	1
In total, 52 cultures	52	15	11	8	1	2	6	1	1	37	8

[1] A *culture series* is a number of cultures made on one day from the wood of a part of a branch or root of a diseased tree.
[2] The figures on the first line under the name of the mold indicate in how many *culture series* the mold named above the column occurs.
[3] The figures which stand further under the molds indicate the *number of cultures* in which the mold named above the column occurs.

Survey of the cultures of the trunk base:

Cultured from	
Bark:	*Cytosporina, Fusarium,* bacteria
Boundary of wood and bark:	Bacteria
Annual rings:	
1921	Bacteria
1920	Nothing
1919	Nothing
1918	Nothing
1917	Bacteria + *Verticillium* sp.
1916	Some discoloration is visible:
	Ramularia, Dematium pullulans?
1915	Nothing
1914	Nothing
1913	Nothing
1912	Bacteria
1911	Nothing
1910	Nothing

The roots, which were very difficult to reach because the ground in which the trees stood was severely hardened so that we have been able to investigate only a few roots, showed no discoloration and in culture yielded no *Cephalosporium-Graphium*, although they did yield *Ramularia*.

In the same way that this thirty-year-old tree was investigated, we now have begun the study of a much older tree in Wageningen.

Further, I have plated out cultures of the wood of very *old* trees, which showed deeper internal discoloration here and there, in order to see whether *Cephalosporium-Graphium* would show up there too, which has succeeded in a few cases. The discoloration occurred, then, in the annual rings [of] 1912, 1913, 1914, or 1915. [See comment on p. 25, p. 6 of original.] The wood for these cultures originated primarily from old trees in Rotterdam (see Plate IV, fig. 5) and was sent to us by the landscape architect there, Mr. D. G. VERVOOREN. For his abundant information and shipments of wood I thank him very much, likewise the technical official of public works at Rotterdam, Mr. A. LEBBINK.

We received much wood from Amsterdam, including also [wood] from trees which showed approximately the same disease picture as the diseased elms through the kind concern of the head of the department [p. 20] of Plantations, Mr. H. C. ZWART, and of the chief superintendent of the Municipal Plantings, Mr. A. VAN LOHUIZEN. Through mediation of these gentlemen we received, among other things, wood of a diseased plane tree and of a diseased poplar. From the wood of neither of these trees could we culture the same molds that come to light with the elm disease.

Finally, I report one shipment from Mr. P. WESTBROEK, Director of Municipal Parks in 's Gravenhage [The Hague], with whom we also maintained a regular exchange of correspondence. From him we received, among other things, wood of a *Pterocarya laevigata*, which had died in 's Gravenhage with symptoms resembling those of the diseased elms. The wood did not show one or more rings of discolored dots but was very dark in color in some branches and roots.

From a portion of that dark wood, too, we have been able to culture *Cephalosporium-Graphium*. This is the **only** case in which we have been able to culture from the wood of a diseased tree of a species other than elm, **the same molds** as those which are characteristic for **the** elm disease. From a later letter from Mr. WESTBROEK it turns out that the tree, *Pterocarya laevigata*, had suffered from illuminating gas in the ground.

Besides *Cephalosporium-Graphium*, there sometimes appear, out of the wood or out of the bark of diseased elms, the molds *Fusarium* sp., *Dematium pullulans (?)*, etc. (see Table 3, page 22).

It is peculiar that, whereas in 1920 the mold *Botrytis* was cultured rather often out of diseased wood, sometimes between bark and wood, in 1921 we **never** succeeded in this. We don't know the reason for that.

So far as the molds *Cephalosporium-Graphium*[1] are concerned (see Plate II, fig. 1, and Plate III, figs. 3 and 4), it already struck us in 1920 that these molds regularly occurred together. It became evident to me this year that *Cephalosporium* under particular circumstances can change over into *Graphium*. If we start out with *Cephalosporium* spores, or with *Graphium* spores, we first get *Cephalosporium* in culture. It then depends on the nutrient medium, whether the mold will keep on growing as *Cephalosporium* or whether in the long run it [p. 21] will convert into *Graphium*. I have never obtained *Graphium* at first; the *Cephalosporium* always precedes it. Under what circumstances *Cephalosporium* converts into *Graphium* I don't yet know. On **solid** nutrient media (stem pieces) I am fairly certain that I will see the *Cephalosporium*-form eventually change over into the *Graphium*-form. Once the *Graphium*-formation has occurred on a nutrient medium in a Petri dish or in a test-tube, then the *Cephalosporium* gradually turns entirely into *Graphium*, and the nutrient medium, if this is not already dark in color, becomes black on the surface. An agar nutrient medium, which dries out, sometimes shows the *Graphium* form at once [but] sometimes not after storage for months.

The *Cephalosporium* mycelium always shows up very rapidly and easily in the cultures from discolored wood. Mostly the white mycelium, with its very regular branchings provided with little conidial heads, is visible the first or second day after the starting of the culture. One need only just stroke across a nutrient medium with diseased wood (as one does, indeed, with bacterial cultures) for [the] *Cephalosporium* mycelium to be visible on the agar a few days later, in the direction of the stripe. It always amazes us,

[1]On the short side branches of the mold threads [=hyphae], the mold *Cephalosporium* forms spores (conidia) which are grouped in little heads.

The mold *Graphium* forms its spores on the tips of threads which stand upright in bundles, approximately as a stable broom [a Dutch stable broom is a round broom, sometimes made of twigs bound together]; the upright-standing bundles of threads are colored dark brown; the spores, which stick to each other, form a small yellow bead (by about 20-fold magnification) on the hyphal bundle.

Spores are propagative organs of the mold.

TABLE 3
TABLE OF THE MOLDS AND BACTERIA CULTURED FROM WOOD OF DISEASED ELM TREES

Places	Number of series of cultures	Names of molds[2]											
		Cephalo-sporium	Gra-phium	Fusa-rium	Botry-tis	Cyto-sporina	Moni-lia	Dema-tium	Phoma	Ramu-laria	Rhizoc-tonia	Verti-cillium	Bac-teria
1920													
Delft	8[1]	4	2	6	3	1	—	2	2	—	—	—	—
Ginniken	1	1	—	1	1	1	—	—	1	—	—	—	—
Hoeven	2	1	—	2	2	—	—	—	—	—	—	—	—
Knijpe	1	—	5	1	—	—	—	—	—	—	—	—	—
Opheusden	6	6	2	5	1	—	—	—	1	3	—	—	—
Oud-Beierland	3	2	2	2	—	—	—	—	—	—	—	—	—
Oudenbosch	2	2	1	2	—	—	—	—	—	2	—	—	—
Rozendaal	2	1	7	2	2	—	—	2	2	—	—	—	1
Rotterdam	7	7	3	2	3	—	—	—	1	—	—	—	—
Schiedam	11	4	3	5	6	—	—	5	—	—	—	—	—
Tiel	3	3	7	2	—	—	—	—	—	—	—	—	—
Tilburg	9	9	—	6	1	—	—	—	5	—	—	1	—
Venlo	1	—	—	1	1	—	—	2	1	1	—	—	—
Wijk bij Duurstede	4	4	4	2	3	—	—	—	—	—	—	—	—
1921													
Amsterdam	2	1	1	1	—	—	—	—	1	2	—	—	1
Delft	1	1	—	1	—	—	—	—	—	—	—	—	—
Dordrecht	1	1	1	—	—	—	—	—	—	—	—	—	1
Haaren	1	1	1	1	—	—	—	—	—	—	—	1	1
Den Haag	2	2	—	—	—	—	1	1	2	—	—	—	1
Joure	3	—	1	3	—	1	—	—	—	—	—	—	—
Nijmegen	1	1	—	—	—	—	—	—	—	—	—	1	1
Opheusden	6	6	6	2	—	—	—	—	—	—	—	—	—
Oudenbosch	3	2	1	1	—	—	—	—	—	1	1	—	—
Renesse	1	1	1	1	—	—	—	—	—	—	—	—	1
Renkum[3]	5	4	3	3	—	1	—	2	—	—	—	—	4
Rotterdam	4	3	1	2	—	—	—	—	—	—	—	—	3
Wageningen	3	2	1	—	—	—	—	—	1	1	—	—	2

[1]See note 1, page 18 [footnote to Table 2].
[2]The figures indicate in how many series of cultures the mold occurred.
[3]Not included here are the cultures cultured from wood of tree 2 of the street in front of the paper factory at Renkum.

therefore, that the mycelium can be cultured out of the discolored wood so easily and so quickly, and yet that we cannot demonstrate it in the wood tissue. See *Microscopic investigation*, page 9 [actually p. 10].

Cultures. Bacteria. Besides fungi, I can often culture bacteria[1] out of the diseased wood and consistently out of the bark. And this was the case as early as 1920, but in 1921 I gave more of my attention to it. The bacteria occur in 2 colors: *white bacteria* and *yellow bacteria*, both on potato agar [PDA].

Before I begin with the identification of these bacteria, I shall await the result of the inoculation experiments done with the bacteria.

Table 3, page 20 [actually p. 22, whole page] gives a survey of the molds and bacteria cultured from diseased elm wood in the years 1920 and 1921.

Infection experiments. As described in the article of January 1921, [p. 23] a number of inoculation experiments were to be done with the fungi gotten out of the diseased elms. These inoculations were begun in June 1920 [and] repeated on several small trees in the summer and autumn of 1920; and regularly, until far into the summer of 1921, new branches of the small trees were inoculated with the molds or bacteria.

Accordingly, I had for each mold, in the beginning of 1921, 1 three-year-old tree, 2 two-year-olds and 2 rooted cuttings (layers). [Unfortunately she does not mention whether these were freshly planted or established young trees.] For the bacteria, with which I did the inoculations only later in the year, I had less inoculation material remaining; thus I could use only 3 small trees for bacterial inoculations. [Presumably here she means three trees for each bacterial isolate.] I made inoculations in the usual way by putting into an incision in the wood a quantity of the mold or the bacterium, by putting the mold just between bark and wood, by putting the mold into the cavity that arose by boring with an increment borer, or by putting the mold into incisions or little drilled holes in the roots. The inoculated spot was closed with wet cotton-wool, around which [was wrapped] Billroth cambric.

One of the two layered-cuttings at our disposal for the particular mold, we put in a large pot with earth and provided the earth abundantly with the mold intended for that group of 5 small trees. Also we put bits of diseased wood into incisions in branches and trunks [twigs and stems]. Naturally the necessary control material was arranged for.

The whole summer through, we noticed nothing [unusual] about the plants. Whereas everywhere in the nation the elms looked bad, were sick, or died, the small trees in the laboratory garden did very well.

Since by chance I needed a twig of elm wood, I discovered early in September 1921 that one of the inoculated branches showed the familiar discoloration.

I then checked a number of the inoculations of the three- and two-year-old elms, and saw the following: some inoculations, done with *Cephalosporium-Graphium*, or *Cephalosporium* and *Graphium* [here, for once, she evidently means each one separately] (the latter is not so easy to obtain separately),

---

[1]It is a very usual phenomenon that bacteria are in diseased plant parts without being the cause of the disease.

clearly showed **the familiar discoloration** in the wood. In **all** cases cultures plated out from this wood that was discolored as a consequence of the inoculation gave *Cephalosporium-Graphium* back again.

We have **not** been able to examine **all** infected branches, since then for the coming summer we [would] have insufficient infected material that could show the dieback process if the mold really causes the disease. The **result of these inoculations therefore is** that *Cephalosporium-* [p. 24] *Graphium* **can** bring about **dark stippling**. Whether the fungi named also can evoke the disease appearance, and thereafter can induce the tree to **die back** still is **not yet demonstrated**.

To date, we have noticed nothing special on the branches inoculated with bits of diseased wood.

Meanwhile a few **other** molds also did give **discoloration** of the wood. I have noticed discolored dots with infections by *Cytosporina*, by *Dematium pullulans* (?), by *Fusarium*, strong discoloration at the pith with *Phoma* and other *Fusarium* species, [and] very strong discoloration, rings of dots, utterly like those of **the** elm disease, with *Botrytis* infections. In **none** of these cases, however, have the cultures made from the infected wood yielded the respective molds *Cytosporina, Dematium pullulans* (?), *Phoma* or *Botrytis*. *Fusarium* I sometimes did get again, on culturing back out of material infected by *Fusarium*, but *Fusarium* is a mold that lives so generally on dead plants, that it never surprises me very much if I get *Fusarium* from diseased or dead plant parts along with other molds.

In **all** cases we have also **always** gotten **bacteria** when reisolating from trees infected with molds, always those which are white and yellow in color on potato agar [PDA], growing exactly like those from diseased elm material.

So far as *Botrytis* inoculations are concerned, these **did** induce **sudden dying** of the branches. Two of the five small trees were killed in the spring, immediately after coming into leaf (a two-year-old and a layered cutting); the inoculations were done in the early spring, before the small trees yet had leaves. We then inoculated a new two-year-old tree on 2 forked branches, at the same distance above the point of forking. The one half of the fork again died at once; the other, somewhat later. In these forks the discolored dots were distinctly observable. On reisolation from the dead material we did **not** get *Botrytis*. (Because of great pressure at the laboratory, I perhaps did these reisolations rather too late, when the wood already was entirely dead.)

From all [of] these **in some respects successful** (?) inoculations, we can **not** determine the **cause** of **the** elm disease with **satisfactory certainty**.

Again I have done new inoculations and shall, in the course of 1922, continue to carry out inoculations; further, the old inoculations will continue to be observed. Also we shall do experiments with grafting diseased twigs on healthy understock in order to [p. 25] see whether we can make the understock sick in this way. I hope, then, at the end of 1922 to be able to say with more certainty whether we really are dealing with a fungal disease or a bacterial disease.

At the same time as these inoculations, I can report the following experiment. In view of the fact that in 1920 we proposed that the dry spring of 1918 might be the cause of **the** elm disease, a large number of roots on one side of each of two three-year-old elm trees were packed in dry cotton and these were wrapped with Billroth cambric. The twigs on the side of the tree where the roots no longer functioned, very quickly showed entirely dried-out foliage. The twigs themselves had brown specks here and there in the wood. I had no opportunity to follow up, whether *Cephalosporium-Graphium* could be cultured out of the brown flecks. These experiments will have to be set up and carried out on a much larger scale, in order to gain any certainty about this phenomenon. For that matter, IF the drought has anything to do with **the** disease, then certainly in the coming summer the elms will surely show the consequences of the drought of the year 1921.

Control. In view of the uncertainty about the cause of **the** disease, **no means of control** can be given as yet. Moreover, it is a question whether it would be feasible, in a large city for example, to control **the** elm disease, even if this were possible.

If a mold or bacterium is the cause, spraying with a fungicide could be tried in a spot where it would be appropriate to use the material (in a city park, for example).

In at least one place in the nation a **carbolineum spraying** has already been done. To date this has **not** been very successful.

As far as the question of the planting of **new trees** is concerned, in place of the missing diseased or dead trees, in any case we should recommend removing the old soil very thoroughly out of the planting holes, and bringing [in] fresh soil, or placing the new elm trees in the place between two old plant holes.

Inasmuch as we cannot imagine that suddenly, from now on, all elms in the nation will no longer thrive, we still advise just planting new elms again, which [procedure], by way of trial, already has happened in various cities. If this will **not** work, [p. 26] that will appear quickly enough and one still can use other tree species.

The choice of **other tree species** is not so easy. This question came to our laboratory very often in the past summer, and consequently we have thought it over seriously. The **elm,** the tree *par excellence* for **city-** and **road-plantings,** is **not** replaceable by another tree. In the cases in which people *still* wanted to convert to other tree planting and took our advice, we named: *maples* (not suitable for all soils), *oaks, beeches, black locusts* (these certainly grow somewhat slowly and the wood, getting older, becomes very brittle, so that large branches can create danger in case of storm), *horse chestnuts,* specifically the double-flowered white, which actually is usable as a city tree only because it makes no fruits (thrown down by the street boys), and yet can also be pruned. In order to grow chestnuts well they shouldn't be pruned, and **a tree which doesn't stand pruning is no city tree.** In the city one must be able to do whatever one wants to do with a tree, and indeed only **the elm** lends itself to this. We heard also of [the] planting of *mountain ash (Sorbus* *aria lutescens*) and *American Linden* (*Tilia americana rubra*).

At the end of the "Control" [add] also the following: An article, "Tree Grippe," that appeared some time ago in the "Nieuwe Rotterdamsche Courant" [a newspaper], in which a report was made in a very unique way about the cause of the elm disease, has raised some [people's] hopes that a "*serum*" would be prepared at WAGENINGEN in order to cure the trees. For those who are to some degree enlightened about the "*serum*" concept, it will be quite evident that there can be no talk of a "*serum*" with trees.

Considering that inquiry was made of us only a short while ago as to that "*serum*," I think that it is useful to report here that with some tree diseases attempts sometimes have been made to control the disease by injecting the tree with particular materials; so far as we know, however, [these attempts] always [have been] without success.

For **the elm disease, no experiments** have been done by us in that direction and therefore also **nothing** has been **achieved**.

Opinions from the practice [from industry practitioners]. It is a mystery to us on what psychic basis, many people who never have had anything to do with the agri-, horti- or forest-culture professions, people who otherwise certainly have not involved themselves with plant diseases, [p. 27] have judged **the** elm disease worthy of their attention. Maybe [that's] because the elm, as a city tree, attracts even more attention than one supposes, and because everyone now realizes that there is something wrong with the elms. Meanwhile, this much is certain: that no plant disease has ever brought [about] so much interest and writing by *unqualified* [people] in the area of plant disease science in the country [in The Netherlands]. The most peculiar opinions, all sorts of strange views, are reported to us in writing or orally, as causes of **the** elm disease. A few of these [suggestions] that have, more-or-less, a right to exist, we shall append here, below.

People very quickly supposed that also other trees: horse-chestnut, blackthorn, hawthorn, plane tree, poplar, were affected by the same disease as the elms. That *large trees* suddenly die is not a usual phenomenon, so if such a mighty tree dies, naturally it catches the eye of those who care. In very many cases it is impossible to state the cause of that sudden death. So far as the cases reported to us are concerned, we consider that often the drought of the year 1921 will have contributed to it, [and has] caused some trees to lose their foliage. However, that does not say that such trees are dead. I think that *many* of the "dead" trees will indeed *leaf out* again next year. Where people asked for help, the advice always is not to cut the trees too quickly, and first to wait, in case the next year the tree perhaps will show that it isn't yet dead.

Commonly I have tried to culture fungi out of [the] so-called diseased material; this usually failed for me, or I got molds that did *not* match the molds of the diseased elms, except in the case of the *Pterocarya laevigata* from The Hague (see page 18) [actually, p. 20], where I *did* culture out *Cephalosporium-Graphium*.

42

According to a few correspondents, one could save the trees by topping them. Upon further investigation, data like this have never turned out to be true. One takes the top out whereupon the roots still have enough strength to support the remaining part of the tree for some time. Soon they [the roots] cannot provide adequate water to *that*, either; the part of the tree now doing service as top [thereupon] shows disease symptoms and is [again] cut away. This goes on this way until the whole crown is cut away and the bare trunk still shows, perhaps, a sprouted bud here and there; quickly this slight remnant of green will also wither and the tree will be dead.

Where I have seen trees [that were] so pruned, they were mostly **[p. 28]** full of elm bark beetles, which beetles, as is known, show up secondarily at diseased trees. I could regularly find the *familiar discoloration* in the wood of the pruned, diseased individuals

With pruned-back individuals one often first sees one-sided withering, while the other side grows through normally. I have observed this a few times myself, yet I have not been able to explain the phenomenon. The withering does not always appear first on the same, windward side.

With unpruned individuals such a one-sided withering is never seen. There the disease usually begins in the top and then works downward, faster or slower.

Sometimes people wrote to us that on one side of the street the trees were all affected, while on the other side of the same road not a single tree as yet showed the disease. [Later it was learned that this effect resulted from spread through root grafts.] Here in Wageningen we have also observed that phenomenon in some degree on a dike, and have considered that we must associate it with the greater drought on that side of the road where the elms withered. We shall have to see what the trees on that dike will do in the coming summer.

Pruned trees can be just as severely affected as unpruned, even though one *sees* the disease in the former only when the tree is in the last stage: the internal wood discoloration immediately gives the evidence of the disease of the tree. That one often *wants* to prune already indicates—in many cases— that such a tree has a sickly appearance and that by pruning one still wants to save something. The roots then seem, as it were, too weak to support the large tree. If one makes the crown somewhat smaller, then the roots can again complete their task.

The constant raising [of surface or grade level] of the ground in the streets [i.e., "fill"], which in the cities they often want to allege as *the* cause of **the** elm disease, certainly has nothing to do with it. I received reports of places where **the** disease was *indeed* present and where it was known for certain that there was no question of fill. I have myself seen elms more than once, which were severely diseased and yet which stood in places where there never was fill.

Further reported to us in the course of 1921, as causes of **the** elm disease, were:

43

—the incompletely burned gasses of the many motors;

—the nowdays-omnipresent electric cables, which, if they break, might allow spreading of electricity through the ground;

—the gasses developed from the bad fuels of the war years (brown coal);

[p. 29]　—the poisonous gasses of the war front[1];

—the river sand from the Maas River, which has streamed through the drainage area of the great battlefields (with that sand the streets were raised).

—the dumping of brine from the ice wagons along the road, or the pouring [discarding, after cooking] of salt potato water in the working-class neighborhoods, over the earth where trees are planted. These two causes can be utterly disregarded of course since the ice-wagon drivers and the "potato-pouring little women," as one of the town supervisors wrote to us, [would have] related to only a few trees and certainly have had no influence on all the diseased trees in the country.

One correspondent from the province of South Limburg would attribute the disease to subsidences in the vicinity of the mines. In that these occur very locally [and] only in Limburg, and the disease is recorded everywhere in the country even where no ground-subsidences occur, we can regard this cause, too, as not valid.

Another correspondent ascribes the disease to the strong bloom of the elms in the year 1920.[2]

Discussion. From the above, the following appears:

We know the [diagnostic] *criteria* for the disease, to wit: sudden withering of all or of a portion of the leaves (at first mostly in the top); the thin twigs are warped toward the middle of the tree, and the tips of the thin twigs are recurved (Plate II, figure 2, and Plate IV); discolored dots [are] in the wood of the most recent annual rings.

We can *demonstrate no mold* in the *discolored woody tissue*; we did succeed in consistently *culturing* the molds *Cephalosporium-* [p. 30] *Graphium* from the *discolored wood*, sometimes *bacteria* as well. From the bark we culture *bacteria*.

A relatively *young* tree *dies* in a shorter or longer time; with *older* trees we do not yet sufficiently know the course of the process.

The *inoculation experiments* give *no adequate certainty* about the cause of the *unknown elm disease*.

---

[1] In the *Comptes rendus des Séances* de l'Academie d'Agriculture de France, Année 1920, volume VI, number 24, there occurs a very short Communication by M. ROGER GRAFFIN, in which is stated that *at the French front the elms* showed themselves very *sensitive* to the effect of the *war gasses*. Nothing is said about the appearance of the elms, discoloration in the wood, etc. It is said, however, that elsewhere too, where the war was less devastating, dead trees nevertheless do occur.

[2] See C. VON TUBEUF: Absterben der Ulmenäste im Sommer 1920 [Dying of the elm branches in the summer of 1920], [which] appeared in *Naturw. Zeitsch. f. Forst und Landwirtschaft*, 18th volume, 1920, pages 228–230: The dying off of elms in Germany in the year 1920 is proposed to be the result of a very strong blossoming of the trees. From the article it does not appear that they [the Germans] are dealing with **the** elm disease, such as we know it in our country. TUBEUF says that the leaves remain present at the end[s] of the twigs, in tufts. That has not been observed in our country.

As yet we can pronounce *no judgement* about the *infectiousness*, or otherwise, of the disease and about a *control material* [measure]. Anyone who wants to be very cautious naturally does not plant elms in the old planting holes; yet it seems to us very desirable, as an experiment here and there, indeed to plant elms again in old planting holes from which the soil has been thoroughly removed with the rest of the [old] roots.

The *drought* of the summer of 1921 will indeed have had some influence on the outward disease appearance of the trees during the summer. From the literature it seems to us that *elms* are *very sensitive* to *drought* and to *heat*.

The question of whether or not to prune the trees is not yet settled, although most directors of municipal park services come more and more to the opinion that it is better, in view of possible infections from external sources *not to prune* at all or to prune *very little*.

Insects, especially *elm bark beetles* occur *secondarily* on weak or diseased trees.

As is known, the disease occurs in cities, in villages, along streets, on squares, in parks, in municipal ornamental plantings, along dikes and national highways, on clay soil, peat soil, sandy soil, in wet places, in dry places; in one word, everywhere that elms grow individuals affected by the *unknown elm disease* occur. The disease arose with a kind of tree that in our country was consistently known as a tree suitable in every sense for street- and road-planting.

The molds and bacteria are not new. Why should these molds or bacteria [in] recent years suddenly have affected so many elms?[1] Notwithstanding our many investigations with [p. 31] older elm woodworkers [e.g., furniture-makers], carriage makers, etc., we did not succeed in getting to know whether the disease already was known earlier. Older professionals are not acquainted with the unique brown rings, and these would certainly have drawn attention if they had already occurred earlier.

From the literature we have learned nothing about such a disease in the elms.

In our opinion, even if we should be dealing at the moment with a fungal or bacterial disease, an external influence having a generalized effect still must have made the trees susceptible to the fungal or bacterial infection. *It is questionable whether we shall ever succeed in discovering this generally working influence.*

Wageningen, January 1922.　　　　　　　Dina Spierenburg, *Phytopathologist at the Plant Disease Service*

[1] It more often happens that a plant disease suddenly arises and then spreads out rather far. This was the case at one time with the oak mildew, known to everyone [a disease imported from North America].

I give, as a recent example, the tomato canker, which suddenly arose in 1919 in our country, in Belgium and nearby Hamburg. Since here we were dealing with greenhouse plants, it was still more surprising that a new disease suddenly arose simultaneously in the greenhouses of various regions. Several years earlier (1907) [p. 31] this disease had caused a lot of damage in different parts of England and later was practically not heard of again.

In such cases, though, one always wonders where such a disease suddenly came from, or by what influence the plants have become susceptible to it. This is naturally very difficult to find out.

Beatrice Schwarz (seated at left, in front of the table) as a graduate student in 1921; with Prof. Dr. Westerdijk (behind table) and Marie P. Löhnis (by window) outside the "Madoera" laboratory building of the Willie Commelin Scholten Phytopathological Laboratory, Baarn.

CHAPTER 3

Marie Beatrice Schwarz

1898-1969

Marie Beatrice Schwarz, known to her colleagues and her many other friends as "Béa," was born in Djakarta, Indonesia (then Batavia in the Dutch East Indies), on July 12, 1898.

In 1919, at age 21, Schwarz was studying plant pathology under Prof. Johanna Westerdijk at the Willie Commelin Scholten Phytopathological Laboratory (WCS), in Baarn, when The Netherlands was thrown into an uproar by the devastating outbreak of the Dutch elm disease (DED). Probably because her doctoral thesis studies already covered trees (willow and peach), her advisor asked her to add the elm problem to her thesis work. The photograph shows her at that time, while she was a graduate student, sitting outside one of the WCS buildings with her advisor and a fellow graduate student, Maria Löhnis. They were Prof. Westerdijk's first two women graduate assistants.

What Schwarz learned about willows and peaches is seldom mentioned. It was for the DED work that she became famous among her fellow scientists. She received her doctorate from the University of Utrecht on April 4, 1922, having discovered the causal fungus of DED and named it *Graphium ulmi* n.sp.

For nearly a decade Schwarz's discovery was the center of controversy. Few scientists outside her own laboratory believed in her work (except Dr. H. W. Wollenweber, in Berlin, and Prof. Dr. Countess von Linden and Lydia Zanneck, both in Bonn), but in 1929 Dr. Christine Buisman showed that Schwarz had been right, after all!

In 1922 Dr. Schwarz returned to Indonesia, the land of her birth, where she accepted a post as plant pathologist at the Agricultural Research Station at Bogor. Her research there included finding strains (in present-day terminology, cultivars) of the ground-nut (peanut), *Arachis hypogaea,* that were resistant to the bacterial wilt disease. Her cultivar Schwarz 21 was generally adopted by growers in Java.

In 1926 Schwarz married T. C. Schol and left research. But after her husband died, during their wartime internment by the Japanese in the 1940s, she returned to The Netherlands with her two sons, both born in Indonesia. She went back to Baarn and joined the Centraalbureau voor Schimmelcultures (the Dutch National Fungus Culture Collection) which at that time was housed in the same building as the WCS. (The buildings of the two institutions now are adjacent on the same property.)

Here Schol-Schwarz conducted research for industry on various groups of fungi, e.g., those that cause fabric deterioration under tropical conditions. She also monographed the genus *Epicoccum* and, in her retirement, she studied *Phialophora*, including the pathogenic "black yeasts." Despite declining health, she continued to work on a second publication about *Phialophora*.

Because of excellence in her scientific accomplishments, in 1968 Schol-Schwarz was made "Officier in de Orde van Oranje-Nassau" (see photograph; the Order of Orange-Nassau is a decoration of honor). She died on July 27, 1969, in Baarn, at the age of 71 years.

In 1970, a memorial was published in the *Netherlands Journal of Plant Pathology* (76:52).

Dr. Beatrice Schol-Schwarz receiving the honor *Officier in de Orde van Oranje-Nassau* from the burgermaster (mayor) of Baarn, 30 April 1969, in the "Madoera" laboratory building of the Willie Commelin Scholten Phytopathological Laboratory.

The Twig Dying of the Elms, Willows, and Peach Trees, A Comparative Pathological Study

Schwarz, Marie Beatrice
1922. *Das zweigsterben der Ulmen, Trauerweiden und Pfirschbäume, ein vergleichend-pathologische Studie.* Utrecht, A. Oosthoek
[Doctor's dissertation, University of Utrecht]*

CHAPTER I. TWIG DIEBACK OF ELMS, WEEPING WILLOWS AND PEACH TREES

[p. 5]   The pure-culture isolation of the fungus from the wood.

I have always disinfected the diseased material externally; for this I shook it 5 to 10 minutes in a 0.1% sublimate [mercuric chloride] solution and then [I] rinsed it thoroughly with pure water. Then it was plated out on cherry-extract agar, using flamed instruments.

This medium serves very well for the isolation of fungi of many kinds, especially wood-inhabiting ones, possibly because it contains many dissolved nutrients in extremely small quantity and the acidity level is rather high [low pH]. The growth of the filamentous fungus is not impeded, but yeasts and bacteria develop less well. It is for this very reason that cherry agar is such an outstanding isolation medium. [They used the slightly acid, Dutch, Mayduke cherry, a rare hybrid between *Prunus avium* L. and *P. cerasus* L.]

Further growing [after isolation] of fungi on cherry agar is possible only in special cases, as for example with the Hymenomycetes. One does best to cultivate the pure-cultured fungus as much as possible on its natural substrate, which has been made germ-free by sterilizing at high temperature. Thus one grows an elm fungus on elm twigs, a willow fungus on willow twigs, a leaf fungus on leaves, etc. This method, however, is not always successful, as will be shown with the peach, and then a better nutrient medium should be sought empirically.

In general the stems of various plants are the best substrates for wood fungi. But they should not be picked too young nor too old, and often it makes no difference whether one uses potato-, tomato-, or lupine-stems.

Occasionally, however, there are among the fungi very strong individual preferences; for example, this is the case with the [p. 6] elm fungus. This [fungus] fruits most beautifully on potato stems and scarcely grows on lupine stems.

*[Chapter 3 (original pages 7–32) was translated by H. M. Heybroek, 9/1985; the Conclusions and part of the Introduction were translated by F. W. Holmes, 9/1986. Chapters that were omitted here: III. Shoot dying and bark scorching of weeping willow, original pages 33–39; IV. Twig dying of peach, original pages 50–67.]

This fact is yet again an indication that only in broad and over-all terms can one give general rules for growing [of fungi] on an artificial substrate, since for each fungus the nutrient medium on which it grows best must be sought anew.

[p. 7] ## CHAPTER II. THE TWIG WITHERING
AND THE VASCULAR DISEASE OF THE ELM

Introduction.

For some years the elm trees which are common in The Netherlands have suffered from a previously unknown disease. The first symptoms appeared in 1919, especially on elms in the province of South Holland. The disease spread rapidly in the following years and reached an epidemic state in 1920 in Rotterdam.

The Director of the city parks, Mr. D. G. Vervooren, asked help from the Director of the Phytopathological Institute "Willie Commelin Scholten," Miss Dr. Johanna Westerdijk, who turned further study over to me.

A certain uneasiness grew concerning this rapidly spreading disease. This is quite understandable since the elm is planted everywhere in cities as a street and park tree. Also, it is almost never absent from the dikes of Holland. It is not only planted as an ornamental tree; its wood is used commercially, too.

The elm is known to be quite smoke resistant, and has here the reputation of placing few demands on the soil. In Germany, to the contrary, it is considered to be one of the most demanding trees, according to Neger (3).

General Appearance of the Disease.

The first external symptoms of the disease are the rapid wilting of the tips of the twigs. These curl and dry up (Plate I, illustration 1). [p. 8] In the year 1920 this occurred from spring into autumn; however, it was particularly frequent in the months of July and August. Although there was an extraordinary amount of rain, one could observe entire tops of trees and large branches dry up suddenly. The leaves dried without preliminary yellowing. The diseased trees in the city parks were immediately pruned back rigorously, and these then budded out again, seemingly healthy and fresh. At this cutting back of the diseased twigs it was evident that the wood had not yet dried, so quickly had the withering process taken place. On the contrary, from a brown ring, which always develops in the wood of the diseased twigs in the vicinity of the cambium, a brown fluid was amply exuded. The curled-up shoot tips were entirely discolored internally.

When I cut into several trees at random, it became plain how widespread the disease was. In a certain part of the city of Rotterdam, in the vicinity of the gas factory, scarcely a tree could be found which did not have the [brown] ring in the wood, even when the tips were not withered.

It is therefore possible that trees are actually diseased without clear external symptoms. It must be mentioned, however, that a trained person can predict from the general appearance of the tree whether a brown ring is to be expected upon cutting into the tree.

Those trees with abnormally small, and only pale green, leaves particularly show, when cut, all the browning of the wood. However, there are just as many cases in which nothing abnormal is visible on the twigs externally, and yet one finds the discolored ring in the wood.

The partial brown discoloration is the primary and only sure symptom of the disease.

The browning appears only in the wood; it expresses itself as a discoloration of the vessel walls. The bark is always perfectly sound. Illustrations 2 and 3 [in Plate I] portray the discoloration that is in a particular annual ring. By counting, one may in a general way—I shall come back to this later—see in which year the disease must have entered into the tree.

The browning was sometimes found in the annual ring of the year 1918.

The disease already has been occurring much longer than they originally thought, but it first showed up over-all in 1919; in the years 1920 and 1921 it even became epidemic.

The ring is not always closed; in Illus. 2, the discolored vessels in the ring are visible separately in the very youngest wood, while the cross-section in Illus. 3 has a much wider, partially discontinuous discoloration in the preceding annual ring; the young wood is healthy.

[p. 9] The spread [extent] of the diseased vessels can best be seen on newly infected twigs which have had the bark removed. The surface of the young wood shows brown longitudinal streaks, as can be seen in Illus. 4, Plate II. In a cross section of such a twig the diseased vessels appear as brown spots. Shortly after the attack, therefore, the brown vascular bundles still occur separately; later the number increases until finally not a single healthy bundle can be found. When one debarks such a twig, the ring of the diseased wood shows dark through the cambium. A cross section of a twig in such a state shows a closed ring of discolored wood.

When one traces the longitudinal streaks in a newly infected diseased twig, the discoloration appears gradually less severe further downward and finally disappears entirely. This phenomenon is noteworthy because it signifies that the disease begins above and proceeds downward. I never saw it any other way.

One can also follow the browning of an older twig upward, and then see how the [brown] ring arises through the anastomoses of the diseased vascular bundles from the one-year-old twigs. On following it [the discoloration] into the latter, one comes finally to the leaves, from which the browning seems to start.

Just as all the branches of a twig come together in the trunk, so the discoloration of the vascular bundles from the twigs can be found gradually to unite until they [the discolored bundles] appear as a single ring in the wood.

51

The ring is not always uniformly developed. One often distinguishes darker and broader [parts], next to weaker [-colored], parts. The former originate from a badly diseased part of the crown.

One may also follow the ring in its downward course. When the trees are very diseased, the discoloration reaches to the farthest branching of the roots, and on the side where the crown has been the worst affected the discoloration is farthest extended.

The Cause of the Disease.

As I have already stated above, the twig withering of the elms in Rotterdam appeared especially severe in a certain part of the city. This was located in the vicinity of the gas factory. The marshy ground there was often filled, and soil aeration was very limited by this [fill].

On removal of the nearly dead trees the ground was found to be extremely acid and rich in *Armillaria mellea* (Vahl) Quél. The supporting poles [p. 10] were entirely covered with fungi. The roots of the elms looked healthy externally, but almost all showed the browning of the wood.

Because of these circumstances people believed [that] the source of the disease could be looked for in the soil, but they overlooked all those cases in which the wood discoloration is limited to the crown. Furthermore it always appeared that the disease spread downward from the twigs. The source is thus to be looked for in the crown itself.

Outwardly, on the diseased and recently killed twigs, one does not see any fungus which could be held as the responsible agent. Locally there appeared on the dead branches numerous red mites, which without doubt do a great deal of damage, but cannot be considered as the direct causal agent because there are also many diseased elms that are free from mites.

It only remained, therefore, to try [to find out] whether an organism could be cultured from the discolored wood. To this [end], I placed bits of the diseased wood, which were sterilized in the usual way, on a cherry agar surface in a petri dish.

At the beginning of the experiments various fungi developed in the petri dish, especially Fusaria. They always establish themselves on long-dead twigs. One can get rid of them, however, if one uses for the isolation, not the molded tips of twigs, but the brown wood from the interior of the twig. In these cases, a certain fungus, and nothing else, always develops from the discolored vessels. The Fusaria and other fungi do not appear further, so they do not occur in the interior of the wood.

I have made many hundreds of isolations and obtained the fungus, but only from the discolored wood, never from the normal-looking [wood]. However, it is immaterial whether the brown wood was from twigs or from roots.

Artificial infection experiments have given evidence that this fungus is the cause of the elm twig death. It is found locally in the wood. Its effect is an affliction of the vascular bundles by which the walls of the vessels are destroyed.

Mycological.

Graphium Ulmi nov. spec.

On the cherry agar surface the elm fungus develops as a white filamentous mycelium which spreads out from the wood in concentric, day-and-night rings. Upon microscopic examination it appears that we are dealing with an *Hyphomycete*. The mycelium is rich in protoplasm and is forked, occasionally branched in a whorl. The hyphae terminate [p. 11] in conidiophores on which many spores are situated, held loosely together by a slime and looking like a small head (Plate II, Illus. 5). They easily disperse in water.

These small spore-heads appear, even under a hand lens, as clear, glistening droplets on the snow-white mycelium. The microscopic appearance is multiform: one may find all the forms reproduced in Fig. 1. The hyphae may be slightly swollen at the tip and bear several larger, sterigma-like forms which pinch off the smaller spores. In a preparation, little bodies of different

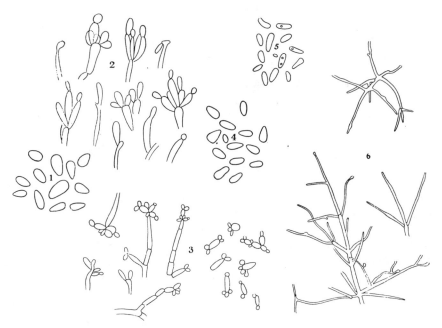

Fig. 1. The different forms of the *Graphium* in culture. 1, Spores of the A-stage grown on cherry agar (×1200). 2, Spore-bearing hyphae of the A-stage grown on cherry agar (×1200). 3, Yeastlike spore development. **B**-stage from cherry juice (×525). 4, Coremial spores (×1200). 5, Spores of A-stage grown on rice (×1200). 6, Hyphae of A-stage grown on rice (×525).

sizes are found side by side, representing spores, sterigmata and all transitional forms between these.

[p. 12] It is this great variability that makes it impossible to classify this form in a particular genus. It is indeed like *Cephalosporium*, but does not fully correspond to it. It is a great hindrance that the fungus cannot be classified in this mycelial stage; for convenience I give this form the name **A** (6 in Fig. 1).

In old cultures round to pear-shaped chlamydospores are produced in the hyphae.

Often the culture in the dish is very sticky, and it appears as though a contamination by bacteria had taken place. Upon microscopic examination one sees in these cases a yeastlike organism. Yeasts frequently appear as contaminants, especially in the isolation of fungi from diseased wood. Therefore, at first I paid no special attention to this, until the culture experiments showed that the **A**-stage under certain circumstances can convert to the yeastlike form, which I will designate further as **B**. I shall consider this question extensively in the following chapter. Here I will only state that the **B**-stage (3 in Fig. 1) does not appear upon isolation, if one dries the pieces of wood with sterile filter paper before putting them on the culture medium.

After only a few days, there show up in the petri dishes, first on the wood and later on the agar itself between the hyphae, dark brown- to black-stemmed coremia with large, clear spore-heads held together by slime. These last can be flushed [washed] off thrice, at most: then no more spores are produced on the same stem.

In the beginning it was presumed that the coremia did not belong to the white-threaded mycelium, yet through culture experiments such was shown to be the case. The variant **A** represents only a lower developmental stage of the coremial fungus of the family Stilbaceae. I have not only germinated many hundreds of coremiospores and always seen the characteristic white **A** develop, but there also resulted, conversely, from hundreds of **A** spores germinated in isolation [i.e., single-spore cultures] the same **A** colonies in which, after the lapse of some time, the coremia were developed. Also in cover-glass cultures, from a single spore the same fungus always resulted, whether from a small head or from the **A**-stage.

In cultures on certain nutrient materials, for example carrots, one may see the transition forms from the usual mycelium to the heads (Plate II, Illus. 7). One finds the peculiar branching of the spore-carrying hyphae from the white mycelium in the head as soon as the superfluous spore mass is removed by pressure. It then turns out that the brown stem, which consists of parallel hyphae with [p. 13] thick-walled, cylindrical cells, becomes gradually clearer toward the top. The hyphae diverge, [and] are then completely hyaline, septate and often forked. At the tips they carry an obovate conidium provided with oil droplets. The difference from the sporophores of the A-stage is that sterigmata are lacking. The spore mass is more homogeneous.

The fungus belongs to the genus *Graphium*, but corresponds to none of the described species. On average, the spores measure 3.25 × 1.71 microns, the extremes are 2–5 × 1–3 microns.

According to the description in Rabenhorst, as far as the size of the spores is concerned, there are a few species of *Graphium* which qualify for closer comparison. These are: *Graphium penicillioides* Corda, with spores of 4–5 × 1.5 microns. The conidiophores, however, are described as being unbranched. Therefore, this species is not identical with ours, because the conidiophores should be regarded as an important character in this genus.

Regarding *Gr. Desmazieri* Sacc., in this case, too, the spores could well agree with those of the elm fungus, as they measure 3–5 × 1.5–2.5 microns. However, the mycelium is smoke colored, and a prominent difference is [the fact] that the sporophores have small bent teeth at the tip.

As a last species in the comparison, *Gr. rigidum* Corda, with spores of 2.5–4 × 1.5–2 microns, could not be discarded completely. Here again the conidiophores, described as unbranched, make it impossible to identify the elm-*Graphium* with this species.

I might mention, furthermore, that from the Botanic Museum in Berlin I got a fungus that was registered as *Graphium rigidum* although it had branched conidiophores. Thus it cannot be assigned to that species. The substrate, on which it occurred as an exsiccata [dried specimen], was decayed wood. I bring up the fungus here, only because it is confoundingly similar to the elm-*Graphium*.

The fact that the *Graphium* that is cultured from diseased elms does not correspond with any described species has led me to name it *Graphium Ulmi*, nov. spec.

The fungus appears in three forms:

1. The mycelium stage, which I designated **A**;
2. The yeastlike stage, which was defined as **B**;
3. The Stilbaceae stage. On the basis of this form I have described the fungus as *Graphium Ulmi* (fig. 8, plate II).

Graphium Ulmi nov. spec.

Description. Coremia in groups, up to 1500 microns high. Stalks unbranched, smooth, up to 1200 microns long and 120 microns thick, constructed of smoke-colored, parallel hyphae that consist of cylindrical, thick-walled [p. 14] cells. The stalks becoming clearer toward the tip, diverging and turning into repeatedly forked, hyaline, septate sporophores that are up to 130 microns long and 2 microns thick. They form a small head which is round, hyaline but in dry condition opaque, yellowish. Up to 350 microns in diameter.

Conidia acrogenous, hyaline, obovate, provided with several oil drops and forming a small head because of slime. The average size is 3.25 × 1.71 microns, the extremes are 2–5 × 1–3 microns. These sizes refer to coremia that are grown on wood.

As an additional form there should be considered a white filamentous mycelium with repeatedly forked conidiophores, which are sometimes slightly swollen at the tip and provided with several sterigma-like structures. The sterigmata are very variable in size and produce by a process of constriction one-celled, hyaline, obovate conidia with several oil-drops. Conidia and sterigmata are kept loosely together by slime and are formed as small heads. On cherry agar the extreme measurements of spores are 2.5–7 × 1–4 microns; the average size is 4 × 2 microns. On various artificial culture media, often in a yeastlike form. In older cultures chlamydospores of varying forms [occur], round to pear-shaped.

As a parasite on *Ulmus* in many places in The Netherlands.

Culture experiments.

In order to get better acquainted with the behavior of *Graphium Ulmi* I started a series of cultures on the most diverse media. The test tubes were uniformly inoculated, using a needle with a loop and a spore emulsion [i.e., suspension] of the **A**-type grown on cherry agar.

The facts are in the Table [Table I]. The differences are very great. Twigs and especially potato stalks turned out to be a very good medium. Lupine stalks are not at all suitable. That is remarkable as these are used readily as a substrate for twig fungi otherwise.

The culture on rice differs from the others by the dark color. It can often be observed that the most intensive color of a fungus develops best on rice. It is not known why this is so. It is not the starch content by itself, as lima bean or oat malt agar do not cause a special color.

Graphium does not grow luxuriantly on a medium that contains only sugar as a carbon source, as can be seen further from the Table.

The difference between cultures on beer wort agar and beer wort gelatin is striking. The latter [cultures] have abundant, filamentous, white mycelium; the former show only a deep yellow, yeastlike mass. It makes another little difference, whether one inoculates with spores [a suspension] in a loop, or with mycelium, at least in the beginning; for in the first case one gets only yeastlike spore heaps [masses], [but] in the second case some scanty, sticky mycelium forms. It appears here again, as mentioned by Appel and Wollenweber (4), that from spores almost only spore-producing cultures grow; from mycelium, almost only mycelium.

In the yeastlike cultures, *Graphium* has a *Tuberculariaceae*-like form, that is, nearly without mycelium. Under microscopic observation, one sees the clumps of yeastlike spores as a mass in which the [p. 17] heavily swollen sterigmata are prominent. Between these can be seen the thinner, scarcely enlarged spores, which are more numerous than the sterigmata. The hyphae occur only sporadically and are composed of short, swollen, cylindrical cells. All sterigmata, spores and hyphae are rich in fat, which shows as larger and smaller droplets.

[on p. 14]

TABLE I
Culture Series of *Graphium ulmi*.

Medium	Color	Mycelium Development	Coremia	Spore Slime
Elm twig	White	Filamentous, sparse, only on cut surface	Here and there formed in bunches. Many on the cut surface of the twig.	Opaque, yellowish
Maple twig	White	Filamentous, sparse, only on cut surfaces	As above	Opaque, yellowish
Potato stalk	—	None	Large and numerous, over entire surface	Hyaline or opaque, yellowish
[p. 15] Lupine stalk	White	Very sparse, only at point of inoculation.	None	—
Bean	Yellowish	Almost no mycelium, culture yeast-like	None	Yellowish spore heaps [clusters, masses]
Carrot (*Daucus carota*)	Yellowish	Culture yeast-like	Numerous imperfectly developed *Graphium* stems; few normal coremia	Hyaline
Apple pieces	White	Turf-like or woolly	Numerous	Hyaline or opaque, yellowish
Meat-broth agar	White	Culture yeast-like; mycelium only on the dried margins, and then forming day-and-night rings	—	—
Cherry agar	White	Mycelium forming day-and-night rings. Culture yeast-like in the center	Numerous	Hyaline or opaque, yellowish
Banana agar	Yellowish	Culture yeast-like	—	Yellowish
Oatmeal agar	White	Sparse	Few	White to opaque yellowish spore heaps
Lima bean agar (*Phaseolus lunatus*)	White	Considerable, forming day-and-night rings	Numerous	Hyaline coremial heads or opaque yellow spore heaps
Rice	Yellowish to dark brown	A little white mycelium; otherwise the culture is yeast-like	Numerous	Hyaline
Beer wort salep agar	Yellowish	Culture yeast-like. Mycelium only on dry margins, sparsely developed	—	Yellowish spore heaps
Beer wort agar	Deep yellow	Culture yeast-like everywhere, also at the margins, and forming growth rings there	—	Deep yellow spore heaps

[*continued on next page*]

57

TABLE I (continued)

Medium	Color	Mycelium Development	Coremia	Spore Slime
[p. 16]				
Beer wort gelatin. No lique-faction	White	Mycelium filamentous and developed in tufts.	Beginning of stem development.	—
Salep agar	White	Mycelium sparsely developed, growing in rings	Rarely developed.	Hyaline or opaque yellow
Mannitol agar	Hyaline	Sparse, yeast-like	—	—
Saccharose agar 4%	Hyaline	Yeast-like, scanty growth	—	—
Cellulose agar	Hyaline	A scarcely visible, yeast-like growth	—	—

The swollen sterigmata, placed in a cover-glass culture, begin again to pinch off yeastlike spores. In Fig. 1, part 3, the different forms are depicted. The conidia can arise on all sides.

The culture on carrots is distinguished by the tufted initial stages of the development of a stalk. Often the stalks do not unfold completely, and one then finds incompletely formed coremia, which represent nice transitional forms from the A-stage to the *Graphium*, as shown in Plate II, illustration 7. The stalk is still loose and not smooth, since everywhere sporophores branch off.

Cellulose agar as such cannot be considered to be a nutrient. The sparse hyphae have developed at the expense of the impurities that occur in the agar and in the cellulose.

Summarizing, I can say: potato stalks, and twigs, are the best media; on these one can obtain the *Graphium* fruiting bodies without difficulty.

Rice- and cherry-agar are just as favorable, although the formation of coremia decreases rather quickly on subculturing. The A-stage, however, persists for a long time on them.

Media like beerwort-, oat malt-, banana-agar and beans I would not recommend because of the yeastlike growth.

[p. 18] In general the size of the fungus spores depends very much on the influence of different nutrient media. In order to be able to express this in clearly measurable data, I have drawn curves showing the length and width of each of 200 spores, from different cultures. Fig. 2 shows width and length of 200 *Graphium* spores from one coremium that grew from diseased wood. The peak for the spore length lies at 3 microns, for the width, at 1.5 microns. The analogous shape of the two curves is striking. In many fungi the width varies in a different way than the length does; therefore only the latter is always used for comparison. In *Graphium Ulmi*, however, this is not the case, and one can use both.

The curves in Fig. 3 are from a coremium formed on potato stalk. On average, the spores are larger; the peaks of length and width are at 4 and 2 microns. The course [shape of the curve] is very regular.

Spores from a coremium, from a culture on cherry agar after isolation from diseased wood, were measured and the curve (Fig. 4) has been compared with a curve of spores from the same culture, but from the [p. 19] much more variable A-stage (Fig. 5). The curves of the coremiospores turn out to conform closely to the coremiospores from the natural substrate (diseased wood). The

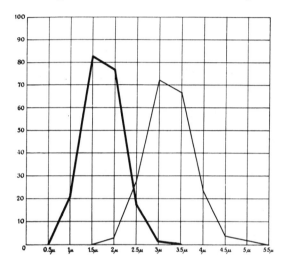

Fig. 2. 200 coremiospores grown on wood. — width, — length. The number of spores is plotted vertically.

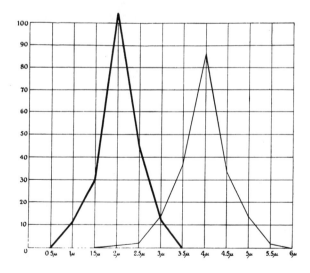

Fig. 3. 200 coremiospores grown on potato stalk. — width, — length. The number of spores is plotted vertically.

59

peaks in this case, too, lie at 3 and 1.5 microns, while those of the A-spores are shifted to 4 and 2 microns. The curves themselves are, by far, not so regular as in the other cases, although length and width diverge again in the same way.

[p. 20]   **Further analysis of the observations in nature.**

As I explained on page 8, the disease is not always externally visible. The browning of the wood is the only criterion of the attack, because it is the

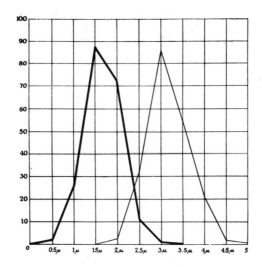

Fig. 4. 200 coremiospores grown on cherry agar. — width, — length. The number of spores is plotted vertically.

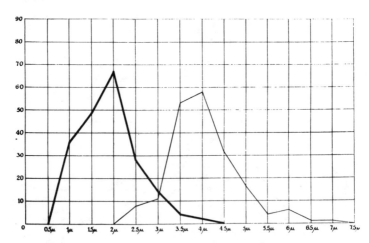

Fig. 5. 200 spores of the A-type, grown on cherry agar. — width, — length. The number of spores is plotted vertically.

discolored vessels that always contain the fungus. This does not occur in the ordinary wood.

Under appropriate conditions, there takes place a spontaneous withering of the twig tips, which curl immediately and confer upon a badly diseased elm its characteristic appearance.

This phenomenon was particularly frequent in the summer of the year 1920, when dying of the elms was so widespread. In the summer of 1921, a different phenomenon was more striking. As early as the months June and July the elms started to discolor yellow and red in the tips of the crown. They had an autumnal appearance and the leaves fell off soon afterwards. The bare twigs were usually dead and dried and had a wide, dark ring, except for the very most recent wood. Also the one-year-old twig tips were not discolored internally; they had developed normally but had died too early, without exhibiting curling. [It seems that she says "one-year-old twigs" sometimes when she means "shoots of the current year" or else she does not distinguish between these two types of twig.] Thus they were not infected, and the early dying, which developed gradually, should therefore be distinguished from the spontaneous and rapidly proceeding withering. It should be ascribed to the abnormal conditions.

I have already pointed out earlier that in the year 1920 the months July and August were marked by excessive rain. During this very period, spontaneous wilting was most common.

The summer of 1921 was unusually dry and it is plausible that the already rather severely reduced water conduction no longer sufficed under these conditions, and the leaves gradually withered. The dying off in the dry summer is therefore purely secondary and the result of an attack that had already taken place much earlier. In contrast to these facts, the curled and internally completely discolored shoot tips represent primary infections. I found them rather seldom in the year 1921. Evidently weather conditions play a large role. Thus, conditions for an acute effect and a quick spread of infection are very favorable in a summer rich in rain.

Primary infections, however, were not fully absent in summer 1921; they just took a less acute course.

I found their marks in the form of individual discolored vascular bundle strands in the one-year-old shoots. These [strands] could be traced upward into a leaf petiole and started in the main vein. This fact is an important phenomenon because it solves at once the question of where the infection takes place in nature. The [p. 21] **leaves, therefore, can come into consideration as the [possible] port of entry for the parasite.**

This process has been imitated experimentally, and confirmed. I put a drop of a spore suspension on leaves which were kept moist in a petri dish. Microscopically, it turned out that the germ-tubes of the spores penetrated the leaf through the stomata. The sporophores with spores emerged again, also through the stomata. Only on the lower leaf surface did I succeed. After 14 days, numerous *Graphium* coremia had formed on the leaf.

61

These experiments should be conducted in a very humid atmosphere. Otherwise they fail. The spores then scarcely germinate and no coremia are formed.

Even though *Graphium Ulmi* makes its way through the leaves, it is not a typical leaf fungus. The infection of a leaf is not visible externally. No leaf spots are formed, the leaf tissue dies gradually, and the vessel browning starts as a weak [faintly visible?] streak, [and] not before the midrib. The fungus can be easily cultivated out of such a leaf.

If one checks the leaves shortly after infection, the browning still has progressed only as far as the leaf petiole. After removal of the leaf, the scar has a completely normal appearance. By careful cutting, one often finds cases, too, in which the browning has passed through the petiole into the twig and has continued downward there. The scar of such a leaf has brown dots at the place of the vascular bundle.

From the above it is easily understandable that mites, in feeding on leaves and petioles, cause damage, in that they facilitate the entry of the parasite through wounds.

In addition, scars of leaves and stipules can be considered as [possible] ports of entry. Indeed this fact is revealed again by the vascular browning, which started in the stipule scars. It is unthinkable that the dry stipules themselves get infected.

The fresh leaf scars can—as has been confirmed in artificial infection experiments—easily be infected. Tyloses are not formed immediately in the torn vessels; the wound surface grows corky later. Therefore, the possibility of a general infection still exists in the autumn, after leaf-fall.

After the autumnal rains, the fallen diseased leaves will undoubtedly bear the small fungal spores, which are easily spread and can infect the fresh scars.

Infection is thus possible nearly the whole year around. It might [p. 22] culminate twice a year, in spring and in fall, after the unfolding and the dropping of the leaves.

It is difficult to say, how far the wood attack spreads in a year. One cannot determine the discoloration externally and measure the spread in the next year, since the twig is destroyed by the cutting.

To be sure, in one-year-old shoots the spread has occurred in a single summer, but the exact moment of infection cannot be proved.

Only by infection experiments can it be determined how quickly the browning proceeds in a year. I am convinced that this depends, in large degree, on external conditions. Unfortunately I have no extensive data on this process, because only by the end of June last year did artificial inoculations turn out to be successful. The few individual data I give below.

The wood-browning apparently has no retarding influence upon the length of the one-year-old twigs, as can be seen from the following figures. In October,

—5 randomly chosen, healthy, one-year-old twigs were, respectively, 17, 22, 25, 40 and 60 cm long;

—5 randomly chosen, bent, dead, one-year-old twigs were, respectively, 30, 40, 41, 60 and 65 cm long.

Anatomical changes evoked by the fungus.

In consequence of the attack, the first anatomical changes in the wood appear as vesicular tyloses in the vessels, which are soon completely filled with them. Later they disappear again and the [vessel] walls are then gum-like, swollen, and discolored brown.

The browning always starts in the vessels; later it spreads out over other elements of the wood, too, which in this stage looks macerated.

To demonstrate the fungus in the wood, cross sections of freshly attacked shoots are required. Between the vessels that are filled with tyloses, there are also a few that are not yet in this stage.

At high magnification occasional thin hyphae can be seen in them, which correspond with the A-mycelium, by their high content of protoplasm and in the type of branching. Illustration 6 in Plate II is a photomicrograph in which the branching is rather clearly visible.

The mycelium in the vessels can be demonstrated best without artificial staining. When the preparation is in glycerin-alcohol for some time, the threads grow further and one can follow them microscopically very [p. 23] nicely by means of a drawing prism [camera lucida]. Ultimately, with some practice, one can get to finding the hyphae sometimes also in the macerated wood, but the wall filaments of the decayed vessels can easily deceive.

Once one has detected the hyphae, it is not difficult to find them in nearly every preparation. However, they never form a large mass that could block the vessels mechanically. It is good to realize in this connection the somewhat mysterious effect of xylem parasites in general.

The effect of *Graphium* is surely a powerful one, as those few thin hyphae are able to provoke such profound disturbances. These must be attributed to a change in the wall.

This vascular disease is different from that caused by *Verticillium albo-atrum* Reinke & Berth. In shoots killed by *Verticillium*, one sometimes finds the interior of the vessels filled with thick hyphae and chlamydospores.

The vessels that are attacked by *Graphium* are on an average larger than the ones normally formed. This fact can even be ascertained with a hand-lens in the ring in a cross section of a stem, in which the brown vessels are still individually visible. By their wide lumina they stand out even more darkly.

The susceptibility of the elm species.

Generally one finds *U. campestris* L. planted everywhere in The Netherlands. [*U. hollandica* 'Belgica' is meant; it was widely called *U. campestris* by park officials at that time.] In addition, in Rotterdam much value is placed on its variety *U. campestris f. monumentalis Rehd.* [*U. carpinifolia* 'Sarniensis'] which is grafted high ["mid-trunk"] or low on *U. campestris* as a rootstock.

U. monumentalis differs from the latter by its pyramidal, compact growth and the rougher bark, which makes the original grafting place recognizable, even in old specimens. Furthermore the leaves are more or less undulating, and they do not fall until November.

No difference in the susceptibility of these varieties has been detected.

A small planting of so-called Candelabrum elms was healthy. Because of their limited number it is undesirable to draw the conclusion that they are less susceptible.

The Candelabrum-elms are fastigiate varieties of *U. montana* Hemsl., such as *U. montana* f. 'Dampieri' Kirchm., and *U. montana* f. *plumosa* [respectively, *U. carpinifolia* 'Dampieri' and *U. glabra* 'Exoniensis'].

Artificial infection experiments.

In the winter of 1920–1921 I performed infection experiments, both on cut twigs placed in water in the lab, and on sprouts in the Cantons Park [Botanic Garden], Baarn, which had remained free of spontaneous infection. Infection was always done by wounds.

In the winter experiments I wounded the twig down to the wood with a sterile knife, lifted a piece of bark, and inserted mycelium with spores from a culture on rice.

In addition I tried to infect the wood via the buds and dormant buds, by sticking a sterile needle through the buds deeply into the wood. Mycelium was thereupon introduced into these wounds, too.

I closed the wounds with raffia, likewise the control experiments, which were carried out exactly the same way but without fungal infection.

It was not possible to elicit an acute wilting and withering by my inoculation experiments. I did succeed in producing wood discolorations which correspond completely with those of spontaneous infections. These phenomena are very clear in longitudinal sections. The *Graphium* mycelium can be cultured out of this discolored wood, too.

Wound closure proceeds normally, starting from the cambium, in that the latter forms slightly swollen callus rims.

In the control experiments and in the cases where infection did not progress, the wound also closed; by cutting into the wood, however, one can see it has not discolored.

I call the cases where a discoloration occurs and *Graphium* develops upon isolation, positive; the others, negative.

In order to study the spread of the mycelium of *Graphium* in wood of different ages, I put spores and mycelium cultured on rice in wounds bored sterilely to different depths. I often used branches many years old.

The experiments were carried out in three ways: a) infections on the tree; b) in a rather large, detached branch system; c) in detached stubs, which were kept in an unheated room. In none of these experiments were there undesired molds.

The spread can be examined by splitting the wood through the drilling wound.

The parasite grows equally fast upward and downward. It does not show a preference for the older or for the younger wood.

[p. 25] By my experiments, Münch's researches could not be confirmed. That is, this investigator [Münch] established the fact that the spread in wood of a large number of fungi is related in a certain way to its [the wood's] content of air and moisture. Let an example be given. When he grows the blue-stain fungus of coniferous wood, *Ceratostomella Pini* Münch, in a moist block of wood, he observes a spread of the fungus as depicted in Fig. 6, part *a*. In dry blocks the fungus grows according to Fig. 6, part *b*. The heartwood, which is always drier than the sapwood, has in the former case an air content which is just sufficient for the fungus; in the latter case the air content in it is too low. [Here her own logic would seem to require that she say the air content was too high, as she has said this wood was too dry.] The mycelium thereupon prefers the sapwood, which contains sufficient air due to the partial drying [and sufficient moisture due to the incompleteness of the drying?].

Sketches 1–3 in Fig. 7 match the experiments mentioned under *a*, which were carried out in the intact tree, while sketches I–V show the progress of the experiments mentioned under *c*, thus those carried out on detached stems. Table II gives the data of infections via deep bore-holes. They invariably turned out positive.

The spread in small wood blocks is less extensive than that in branches in the tree. This [fact] indicates a high degree of parasitism. The result is contrary to what, according to Münch, would be expected.

The bark at the side of the wound is slightly discolored. Apparently the parasite grows in the tissues that are killed by the wounding. In experiments 1–3 the bark was too thin to let any discoloration be seen.

[p. 27] Concerning radial spread, the sketches show that everywhere the fungus was found [to be] deeper in the wood than it was originally inoculated. This is an important fact; it means that one cannot determine with absolute certainty in what year the disease must have occurred for the first time.

Fig. 6: Sketch according to Münch. a) Block of wood with high moisture content. b) Block of wood with reduced moisture content.

The artificial infection further shows that the fungus grows with the same speed upward as downward. This appears to be in contradiction to my opinion about the normal spread from the upper parts to the lower ones. However, one must not forget that these infections [the author often says "infection" where a present-day pathologist would say "inoculation"] all took place in branches many years old, while in nature only infections in the younger parts play a role. I can only say that in all observations in nature a spread away from the growing parts was always proven. Last summer, many dormant buds in several trees grew out as late as August, because of the drought. I checked many of the young shoots for the browning, but found them always perfectly healthy, even though the underlying wood of the stem was completely discolored.

In nature, therefore, the parasite always grows toward the older parts.

In the summer, I used younger tissues for inoculation, e.g., shoot tips, leaf-blades, petioles, and leaf-scars. I chose strong sprouts which had remained free of natural infection, even though there were heavily diseased trees in the same garden.

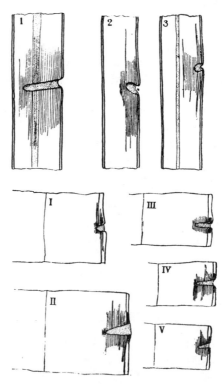

Fig. 7. Sketch of artificial *Graphium* infections in older branches. 1–3) Experiments in the standing tree. I–V) Experiments in detached blocks.

This time I inoculated with a loop with a spore suspension, and always into wounds. In the bark, these varied from narrow, deep cuts penetrating into the wood, to wide wounds reaching only to the cambium. All were grown over with callus later.

The leaf inoculations could not be closed with raffia; in the other cases this was standard practice.

On leaves, the epidermis was locally wounded by abrasion. The infection progressed if it were performed close to the midrib. The wounded leaf tissue simply died, but in the vascular bundles of the midrib brown streaks were visible, which, when the experiment was checked (sacrificed), were still restricted to the petiole. The fungus is easily cultured from these parts.

The experiments suffered great damage from the drought; a whole series had to be excluded from consideration, because the tree in question dried up completely. There were no symptoms of *Graphium* in this case. Many leaves dropped; unfortunately, inoculated leaves also were among them.

[p. 26]

TABLE II
Infection Experiments on Twigs of Several Years' Growth.

Experiment Number	Duration of Experiment	Infection Material	Twig Diam. in cm.	Wound Depth in cm.	Result	Radial Discoloration in cm.	Longitudinal Discoloration in cm. Upward	Downward
1	Jan. 12 to Sep. 15	*	1.7	1.4	Pos.	1.6	1.8 in medulla	2.8 in young wood
2	↓	*	1.1	0.35	Pos.	0.55	2	2.5
3	↓	*	1.5	0.25	Pos.	0.65	1.4 in young wood	1.8
4	Jan. 12 to Oct. 20	*	—	1.2	Pos.	—	4	4
5	↓	*	—	0.2	Pos.	—	1.5	1.5
6	Jan. 11 to Sep. 15	*	1.5	0.5	Pos.	—	9.5	4
7	↓	*	0.6	0.2	Pos.	—	0.6	1
8	↓	*	0.7	0.3	Pos.	—	30	3
9	Jan. 11 to July 10	*	4.5	0.25	Pos.	0.75	0.9 in young wood	0.6 in bark
10	↓	*	5.5	1	Pos.	1.1	1	1
11	↓	*	3.5	0.6	Pos.	1.1	0.3	0.2
12	↓	*	4.5	0.6	Pos.	0.65	0.7	0.6
13	↓	*	4.5	0.4	Pos.	0.45	0.6	0.4

*["Infection material" for all 13 cases in this column reads: "Mycelium and spores from a pure culture on rice."]

Remarks:
Experiments 1–5 were made on twigs on the tree; 1–3 are as depicted in sketches 1–3 in Fig. 7.
Experiments 6–8 took place on a large branch system that was placed in water. It bloomed in February; was dried out in March.
In number 6 the discoloration reached downward as far as the cut surface. In all cases the diseased vessels lay just under the surface.
Numbers 9–13 were made on detached blocks [pieces of stem]. Sketches I–V in Figure 7 relate to these.

I inoculated leaf scars immediately after the removal of the leaf. They yielded very fine results.

[p. 28] Tables III and IV contain the result of the experiments on comparable sprouts. It appeared clearly that inoculations performed in summer are more successful than those in winter.

The positive artificial inoculations have proved beyond doubt that *Graphium* is the cause of the elm wood browning. The characteristic external symptoms, caused by a rapid withering, however, have never been produced artificially. Two causes for this come under consideration: a) the abnormal weather; b) the fact that sprouts were used.

Sprouts are undoubtedly valuable material for inoculations, as they either do not or, in any case, with great difficulty become spontaneously diseased. On the one hand, there is the advantage of having no confusion between

TABLE III
Infection Experiments on Sprouts

Experiment Number	Duration of Experiment	Infection Material [Inoculum]	Place of Wound	Result	Remarks
14	Nov. 17, '20 to Aug. 12, '21	Mycelium and spores from cultures on rice. Isolated from roots.	Bark	Positive	Controls all negative.
15			Dormant bud	Positive	Controls all negative.
16			Dormant bud	Negative	Controls all negative.
17		The same, but isolated from the stem.	Dormant bud	Positive	Controls all negative.
18			Dormant bud	Negative	Controls all negative.
19			Bud	Negative	Controls all negative.
20			Bark	Negative	Controls all negative.
21	Dec. 22, '20 to Aug. 12, '21	The same as in experiments 17–20.	Bud	Negative	Controls all negative.
22			Bud	Negative	Controls all negative.
23			Bud	Positive	Controls all negative.
24			Bud	Positive	Controls all negative. In Expt. 24 the discoloration also developed 0.5 cm. upward.
25			Bud	Negative	Controls all negative.
26			Bud	Negative	Controls all negative.
27			Bud	Negative	Controls all negative.
28	Jan. 10, '20 to Aug. 12, '21		Bud	Negative	Controls all negative.
29			Bud	Negative	Controls all negative.
30			Bud	Negative	Controls all negative.

TABLE IV
Infection Experiments on Sprouts

Experiment Number	Length of Experiment	Infection Material [Inoculum]	Place of Wound	Result	Spread of Discoloration in cm.	Remarks
31–32	May 28 to Aug. 5	Spore suspension from culture in A-stage grown on cherry agar	2 × Bark	2 × Pos.		All checks negative everywhere. The spread measured only in a few cases.
33			Bark	Pos.	1	
34			Bark	Neg.		
35			Leaf blade	Pos.		
36			Leaf blade	—		Leaf fallen; not checked
37–39			3 × Bark	3 × Pos.		
40			Bark	Neg.		
41			Leaf blade	Pos.		
42–43			2 × Leaf blade	—		Leaves fallen.
44			Leaf blade	Pos.		
45–47			3 × Bark	3 × Pos.		
48–49			2 × Leaf petiole	2 × Pos.		
50			Bark	Doubtful		
51	July 13 to Aug. 5		Bark	Doubtful		The experiment was halted as early as after 3 weeks on account of the great drought.
52–53			2 × Leaf blade	2 × Neg.		
54–56			3 × Leaf scar	3 × Pos.	0.3	
57–60	Sep. 4 to Jan. 23		4 × Leaf scar	4 × Pos.		In experiments 57–76, the bark wounds extended to the cambium only.
61–62			2 × Leaf scar	2 × Pos.	0.75; 0.5	
63		Suspension of coremial spores.	Bark	Pos.	2	
64	Sep. 4 to Nov. 19		Leaf scar	Pos.		
65–67	Sep. 4 to Jan. 23		3 × Bark	3 × Pos.	0.75; 1; 1	
68–69			2 × Bark	2 × Neg.		
70–71			2 × Leaf scar	2 × Neg.		
72 – 74			3 × Leaf scar	3 × Pos.	0.5; 0.2; 0.75	
75–76	Sep. 4 to Nov. 10		2 × Bark	2 × Pos.		

artificial and natural infection. On the other hand, this low susceptibility makes artificial inoculation much more difficult. I chose the sprouts [p. 29] also because they are so easy to observe, and layered plants were not available to me.

In order to find out whether *Graphium* generally can easily penetrate living plant parts, it should be added that I put *Graphium*, without wounding, on freshly cut stalks of tomato and lupine, and that I infected the cut surface of a potato with the mycelium. The experiments were carried out in a humid atmosphere, in petri dishes. The fungus did not penetrate in any instance; the characteristic discoloration did not appear.

[p. 30]   **Weather influences and possibilities for control.**

In regard to weather I will make only some general remarks. Evidently it plays a big role with tree diseases, too. I regret [that] at present I cannot yet express the relation between the weather and the elm disease numerically, but it will be interesting to consider this in the course of time.

At the present state of our biological knowledge a growing tendency can be seen which tries to analyze the interrelationships of living beings with the help of exact numerical data. The analysis of the special interrelationships that are of use for agriculture and forestry—thus, [analysis] of the struggle between our cultivated plants and their enemies—should, of course, be first on the list. The weather is of extraordinarily great importance in this respect, as is also shown for the animal enemies of our cultivated plants in the recently published and fundamental work of Carl Börner (5).

The Phenological Society, recently founded in our country, undoubtedly can supplement the meteorological observations with its data and be a strong help in the analysis of the effects of the weather. It fills a great need.

As the further course of the elm disease under the influence of the weather cannot yet be predicted, I do not think it advisable to remove the diseased trees immediately.

If new infections do not take place in considerable numbers and area, [then] I believe the trees will recover externally very well and be able to live on with the help of the young wood. For one need not fear a radial spread of the fungus in the direction of the new wood, in conformity with [in view of] the fact that, in the natural course of events, the parasite always tends to move away from the younger parts.

Control of the disease is, due to its nature, hardly possible. Prevention of infection could be tried by the spraying of some fungicide immediately after leafing out, thus before infection occurs. In certain cases, e.g., where beautiful park trees are concerned, perhaps this method may lead to success. The right time for spraying should be determined exactly.

One can counter the spread of the browning into the stem by effecting an early [prompt] removal of newly infected parts. One need attribute no great importance to natural infection of the older wood by way of the cut surfaces. The pruning therefore will do no harm.

Literature.

In the year 1921 a small publication appeared by Miss D. Spierenburg (6) on the elm disease. This had, however, wholly the character of a preliminary communication.

Although they have no direct connection with the elm disease, two detailed works by Münch (7,8) come into consideration for a short discussion. In connection with my researches it is remarkable that the genus *Graphium* has among its representatives two other species that are considered as wood parasites, albeit on felled wood. They are the secondary fruiting forms of two blue stain fungi.

Blue stain of coniferous wood is, as we know, a common and important phenomenon, about which we have the works of Münch (7,8) from the years 1907 and 1908. The only thing that interests us in these papers is that two of the fungi that are considered as the cause produce as an imperfect stage a *Graphium* that is not further identified. In addition they form a mycelium that by its variability shows a great resemblance to the mycelial stage **A** of *Graphium Ulmi*. They are the Ascomycetes *Ceratostomella Piceae* Münch and *C. cana* Münch.

Although he describes and depicts the *Graphium* belonging to *C. Piceae* with branched conidiophores, he still thinks it possible that this coremial fungus be identical with penicillioides. This is yet a further proof that the systematic treatment of the genus *Graphium* is very deficient.

The blue stain fungi do not penetrate into the interior of fresh living wood; they do not attack the wood substance to a great degree, but live only on the contents of the parenchyma cells. They are thus entirely different from *G. Ulmi* in their effect and [are] less potent.

The occurrence of an ascus fruiting form indicates the possibility that the fungus of the elm disease is able to develop an additional fruiting form, an ascus form.

C. von Tubeuf (9) ascribes the elm dying in Germany in the years 1918 and 1920 to the excessive flowering. He understands the phenomenon as purely physiological, thus as a consequence of the depletion of nutrients.

Although I am not certain, because he does not describe the symptoms in detail, I think it is not impossible that the same *Graphium* disease is in question here.

Summary of the disease phenomena.

The effect of *Graphium Ulmi* expresses itself as a vascular browning, which occurs in the wood locally and which is the result of a destruction of the vessel walls.

The infection takes place by way of the leaves (stomata, wounds) and the leaf scars and stipule scars.

The result of an infection, beyond the browning which always occurs, can be twofold:

a) A rapid wilting of the shoot tips, which curl immediately, takes place. This causes the acute disease syndrome.

b) The effect is not visible externally. Only a relatively long time afterwards do the diseased twigs gradually die.

Both phenomena are greatly influenced by external conditions.

The disease always spreads from the points of infection toward the older parts. In the artificial inoculations the parasite shows no preference and [so] here [it] spreads in all directions.

[End of original p. 32 and of Chapter II]

[p. 68]   CHAPTER V. GENERAL CONCLUSIONS.

If we compare the twig death of the elm, weeping willow, and peach trees with each other, great differences appear.

The elm disease can immediately be set apart from both of the others. Here [we] are dealing with a single, hitherto unknown, fungus, *Graphium Ulmi*, as pathogen. This probably is adapted solely to the elm; thus it has achieved a very high level of parasitism.

Although the organism makes its way through the leaves, still it does not manifest itself in its attack as a true leaf fungus, which destroys the leaf tissue. The effect is a restricted but powerful wood injury, by which the very thin hyphae, only rarely visible, decompose the vessel walls and evidently thereby cause a profound derangement in the provision of water. They do not plug the wood mechanically, but perhaps it is due to the plant's own reaction, which instantly develops tyloses, that the water conduction is arrested.

Here one can—following van der Lek (26)—speak of a tracheomycosis, since it is a typical vascular disease, without thereby casting substantially more light on the phenomenon.

This example enriches the analysis of the twig dying, as I have performed it in the first section, with a new phenomenon, to wit, the localized vascular attack. This differs from wood-deterioration or decay, which is caused by *Hymenomycetes* and which is very energetic [affects the whole tissue], whereas attack by vascular parasites is confined to certain vascular bundles, at least at first.

With weeping willow and peach, several organisms come under consideration as causes of death. However, they all live ultimately as bark parasites, which only secondarily can move over into the wood. Each one is itself very abundant, and most are not at all adapted to a single host [p. 69] plant. The large number shows that in our climate both host plants are in unfavorable circumstances, in which they are subject to the attack of diverse fungal species. Moreover, the weeping variety of the willow is in itself a weaker one.

With the willow, the general reaction to a parasitic attack is the browning of the leaves and the blackening of the bark.

With the peach, as result of the bark attack, the disrupted cambium begins to exude gum.

The most important parasites of the willow and the peach, namely *Fusicladium* and *Monilia*, are fundamentally different in their primary attack. The first begins as a true leaf fungus and only moves secondarily onto the twig. This picture thus is comparable to that evoked by *Gloeosporium nervisequum* on the plane tree [sycamore?] (page 4). *Monilia* uses the blossoms and the very young fruits to gain entrance to the bark.

Fusicladium can develop only bark cankers on the twigs—also deep ones, penetrating as far as the wood; it then dies off and entry is provided for many other fungi.

According to statements of other authors, *Monilia* can stay alive in the places of the bark scorch it originally caused, and in the spring provide new infection material in the form of newly developed conidia. In the twig the mycelium spreads further and the bark scorch develops into a canker by evoking multiple callus edges.

The death of the willow and peach twigs first takes place when the fungus has spread around the whole twig, thus as result of girdling, developing inward from [the] outside. With the twig death of elm the bark stands in no such relation to the *Graphium* attack. The same result—killed twigs—thus comes into being with the elm in a way [that is] essentially different from [that] with the peach and the willow.

As far as the spread of the disease is concerned, I hope in each Chapter to have emphasized sufficiently what an unusually large role the weather plays in this regard. This fact is already generally recognized in phytopathology with diseases of other than woody plants.

On perennial plants, as our trees are, it is possible that, under favorable conditions in the next growing season, they can overcome rather well the damage of the preceding year, especially if it is a matter of foliar injury.

One need not be too afraid of the *Graphium* disease up to now. [It is] only if the parasite finds favorable conditions [p. 70] for its development over several years, [that] the damage will increase every time.

In any case one cannot get rid of the parasite in the trees once they are attacked, since it evidently is still alive in the decomposed wood.

The weather has a differing influence on the different plants and their interrelationships with the pathogens. This fact can be illustrated by a comparison of the elm and the weeping willow. They were both very diseased in the rain-drenched year 1920. The willow, however, in large part recovered in the past year 1921; the summer drought of the same [year] was very injurious to the elms with their partially destroyed wood.

In the study of the mycology of the twig diebacks, the conflict with systematics returned time and again. Phytopathology is assigned not only the task of studying the disease phenomena; it often has to fight [live] with a very

73

defectivelyreedited and outmoded mycology, which frequently is based only on incompletely provided spore sizes. The revision of fungal taxonomy through culture- and infection-experiments has already been taken strongly in hand by the phytopathologists—I need hardly mention Appel and Wollenweber (4) with their *Fusarium* monograph, as the founders of this [approach]. There are several more examples, I shall not itemize them all. Up to now they do not yet relate, of course, to the wood parasite [i.e., to *Graphium ulmi*], but out of this [present] work [it] is sufficiently obvious that in this regard, too, a revision of the taxonomy is not only necessary but, on the whole, inevitable. [End of dissertation]

[p. 73]   **BIBLIOGRAPHY.**

[The only items given here are those cited in the present translation.]

(3) Neger, F. W. Die Laubhölzer. 1920, 47.

(4) Appel, O. & Wollenweber, H. W. Grundlagen einer Monographie der Gattung Fusarium (Link). Arb. Kaiserl. Biol. Anst. L. u. F. w. 8, 1910, 1-207.

(5) Börner, C; Blunck, H.; Speyer, W.; Dampf, A. Beiträge zur Kenntnis vom Massenwechsel (Gradation) schädlicher Insekten. Arb. Biol. Reichsanst. L. u. F. w. Bd. X, 1921, 405-466.

(6) Spierenburg, D. Een onbekende ziekte in de iepen. Versl. Meded. Phyt. D. Wageningen No. 18, 3-10. [1921]

(7,8) Münch, E. Die Blaufäule des Nadelholzes. Naturw. Ztschr. L. u. F. w. 5, 1907, 531-573; 6, 1908, 32-47; 297-323.

(9) Tubeuf, C. Freiherr v. Absterben der Ulmenäste im Sommer 1920. Naturw. Ztschr. L. und F. w. 18, 1920, 228-230.

(26) Lek, H. A. A. v. d. Onderzoekingen over Tracheomycosen. Meded. Landbouwhoogeschool XV, 1918, 1-45.

[p. (74)]   **POSTSCRIPT.**

[This part of the text was in Dutch.]

Having come to the end of this research, I want to thank all who were helpful to me in the performance thereof.

By means of Mr. *Vervooren* I could get from Rotterdam as much elm material as I needed.

The completion of the drawings and photos took place with great care by Mr. *A. de Bouter.*

I have Mr. *van Luyk* to thank that the fungal taxonomy, so confusing for a beginner, lost much [of its] horror for me. His rich experience frequently came to my benefit and I could freely draw upon his herbarium.

I also want to point out that I carried out this research as guest of the Willie Commelin Scholten Foundation. At the moment mostly students at the Utrecht National University profit from this Laboratory, which is open to everyone who wants to work there. The Willie Commelin Scholten Laboratory gives them that opportunity, and—it stands alone in this, up to now—[the opportunity] to learn to treasure at its full value the Canton Park in Baarn, the Botanical Garden of the National University of Utrecht. The fact that one can study the material from the Garden directly, in the [neighboring] Laboratory I have discovered as an especially favorable factor.

With a thankful feeling I think about the nice way in which I could get, by means of Mr. *Goossen*, Botanical Garden Curator, everything that I needed for my research.

I feel extraordinarily indebted to Professor *Westerdijk*, and not only for the interest in my work and the guidance experienced therewith. It is most of all the special atmosphere which she can create, which will leave a permanent impression on me.

Baarn, Feb. 1922

ILLUSTRATIONS

[All seven plate illustrations were photographs. The other five were about diseases of willow and peach.]

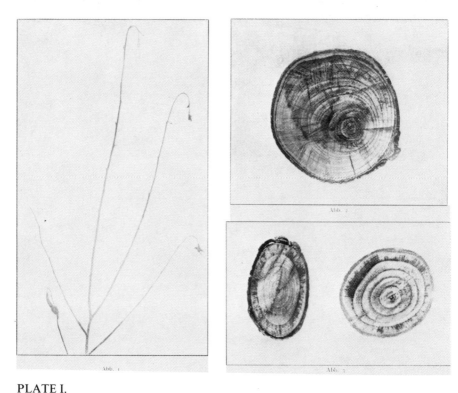

PLATE I.
1, Acutely wilted and curled elm shoot tips. 2, *Graphium* injury in the youngest annual ring. The brown vessels [are] visible singly. 3, *Graphium* injury in elm wood. The ring is not closed. The broad discolored parts correspond with several diseased side branches.

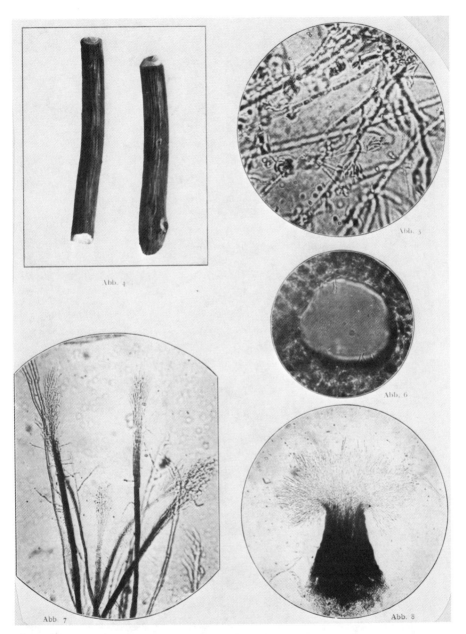

PLATE II.

4, Debarked elm branches. *Graphium* injury in the form of dark longitudinal streaks in the most recent wood. 5, *Graphium* mycelium (**A**-stage) grown on cherry agar. Enlarged 700×. 6, Wood vessel with branched *Graphium* hyphae. Enlarged 700×. 7, Incompletely developed *Graphium* coremia grown on carrots. Enlarged 300×. 8, *Graphium* coremium. Enlarged 200×.

Prof. Dr. Johanna Westerdijk, in the Centraalbureau voor Schimmelcultures, Baarn (seated, examining a culture in an Erlenmeyer flask) in 1928. The woman standing is Catharina Cool.

CHAPTER 4

Johanna Westerdijk

1883-1961

Johanna Westerdijk was known to many as "Hans," which is normally a boy's name; the nickname was given her by her father. She was born January 4, 1883, in Amsterdam, entered primary school in 1888, and began studies at the Girls' High School in 1896. She had one sister. Her mother, from France, became deaf early in life. Her father was a country doctor (general practitioner) with a Groningen background. The family lived on the outskirts of the city, along the beautiful Amstel River. From childhood on, Johanna was remarkable for her independence of thought and action. She gave much credit to her parents for her multiple interests.

Westerdijk's original plan, to become a professional pianist (her mother's sister Johanna, for whom she was named, taught music), was blocked by neuritis in her arm, so she turned to botany. However, because she was a girl, she was refused as graduate student by Prof. Hugo de Vries at the University of Amsterdam. Through a friend, C. J. J. van Hall, however, she became assistant to Prof. Ritzema Bos, the first director of the Willie Commelin Scholten Phytopathological Laboratory (WCS), which had been established in 1895 (and did not move from Amsterdam to the large house in Baarn until 1921).

At the age of 22, after completing the equivalent of a Master's degree under Ritzema Bos in 1904, Westerdijk went to study abroad (as her father had done)—to München under Prof. K. Goebel and then to Zürich under Prof. H. Schinz. There, in 1906, she received her doctor's degree, with a thesis on the regeneration of mosses, and promptly returned to The Netherlands.

As Westerdijk arrived, Ritzema Bos was just leaving the Willie Commelin Scholten Laboratory to join the Agricultural University faculty at Wageningen, and Westerdijk was offered the position of director. She embarked on her new duties on March 15, 1906, at Roemer Visscher Street, Amsterdam.

It speaks something for the powerful character of this remarkable woman that, at the age of only 23, with no time for any professional career accomplishments since her doctoral studies, and with professional specialization in a wholly different aspect of botany, Westerdijk was offered the directorship of the Willie Commelin Scholten Phytopathological Laboratory. The suggestion has been made that her appointment may have been influenced by the fact that Mrs. Scholten, the mother of the deceased young botanist in whose memory the WCS was created, was a fiery feminist.

Only a year later (1907) Prof. F. A. F. C. Went, of the University of Utrecht, asked Westerdijk to care for 80 pure cultures of fungi, a collection established in Leiden in 1903. This collection she eventually called the Centraalbureau voor Schimmelcultures (CBS), and it became a separate foundation under the Dutch Academy of Sciences. Today its staff maintains and studies some 30,000 isolates and strains of about 8,000 species of fungi, from all over the world. Westerdijk was one of the founders of the Dutch Mycological Society (Oct. 17, 1908) and was its first secretary (until Sept. 1912).

Westerdijk's interests were worldwide. In 1913 she visited the Dutch East Indies (now Indonesia), where she studied diseases of tropical crops. She went on to Japan just before the outbreak of World War I. She continued eastward to the United States, where she supported herself by lecturing at various universities. Here she again encountered prejudices against women in science, a plague still rife in many parts of the world.

In 1917 Westerdijk was appointed professor in plant pathology at the University of Utrecht and, in 1930, at request of her students, also professor at the University of Amsterdam. Her clear lectures attracted many to join the field of plant pathology. She was a prominent member of the Dutch Society of Women with Academic Education and was the first woman full professor—of **any** subject whatever—in The Netherlands' history. Of the 54 students who completed doctor's degrees under her guidance, 25 (46.3%) were women, an unheard-of percentage both then and now!

As director of the WCS, Westerdijk was active not only with fungal diseases but also with physiogenic diseases (e.g., effects of acid soil on oats), viral diseases (e.g., mosaic in greenhouse tomatoes), and bacterial diseases (e.g., poplar canker). She also gave careful attention to problems of control measures, as posed by growers, including chemical treatments for seed disinfestation. Among tree diseases, she studied those caused by *Amillaria mellea, Fomes annosus, Stereum purpureum, Gloeosporium* species, and *Nectria galligena*, and—beginning in 1919—the new epidemic among elms. Today this is known as "Dutch elm disease" (DED) mainly because of the excellent Dutch research on it in the 1920s, which was largely promoted and directed by Westerdijk.

To study DED, Westerdijk enlisted the help of the Dutch Heath Society (the Dutch Heathland Reclamation Society, founded in 1888 after an early Danish example). Then she helped to organize the Elm Disease Committee. Funds were solicited from every Dutch town and city. Under Westerdijk's direction, M. B. Schwarz discovered and named *Graphium ulmi* as the true cause of the disease; C. J. Buisman confirmed this, identified the perithecial stage, started selection for resistant elms, and cooperated with J. J. Fransen (in Wageningen) on studies of elm bark beetle transmission of the fungus; J. C. Went initiated breeding for DED resistance; M. S. J. Ledeboer studied the physiology of the pathogen; and S. Broekhuizen studied the elm's physiological and morphological reactions to DED.

In 1917 Westerdijk proposed a book to group plant diseases according to symptom types, and in 1919 she published an extensive article to this effect with Prof. Dr. Otto Appel, for many years Director of the Biologische Reichsanstalt, in the Dahlem suburb of Berlin. She had a high command of the English, French, German, and Spanish languages.

Westerdijk's zest for life, her enthusiasm for work, and her strongly expressed opinions exerted a powerful influence on the students and others with whom she worked. Occasionally resentments arose, but generally there was great admiration. Above the door of the Madoera annex, rebuilt in 1929 to supplement the WCS, is a motto that fits her well: "Werken en feesten vormt schone geesten" [Working and feasting develop fine minds]. She was fond of waltzing. She liked and created a free and happy atmosphere.

Her example and ideas about life provided a rare inspiration to her women students of phytopathology. However, it should be added that her laboratory was not a female bastion, hostile to men. The fact that it had a relatively high female population shows, instead, that other laboratories and other professors were less inspiring for female students (or possibly less inclined to encourage them in scientific careers).

For 20 years, Westerdijk was coeditor of the journal *Tijdschrift over Plantenziekten*. She was president of the International Federation of University Women, presiding in 1937 at Krakow. In 1938 she visited South Africa. From 1945 to 1951 she was President of the Netherlands Phytopathological Society. She was made an honorary member of the Dutch Mycological Society at its golden jubilee in 1958. She was a Knight in the Order of the Dutch Lion, an elected member of the Royal Netherlands Academy of Sciences, a Fellow of the Linnean Society, an honorary doctor from the University of Uppsala (1957) and from Liebig University (1958), a Knight of the Order of Santiago da Espada (in Portugal), and the first recipient of the Otto Appel Medal at Heidelberg (1958).

At Prof. Westerdijk's last lecture, to 500 persons on Nov. 22, 1952, upon her retirement, a fund was established in her name for the advancement of plant pathology. She died on Wednesday, November 15, 1961, at the age of 78, in her apartment in the Willie Commelin Scholten Laboratory building, which had been her home for more than 40 years.

Among the many sources of published memorials or biographical information are the following:

Antonie van Leeuwenhoek 28(1):1-4, 1962.
Coolia 9(1-2):3-8. (Apr. 1962; A. Jaarsveld)
Feminine Vignettes of The Netherlands B-5(20):1-4. (Oct. 15,1952)
Jaarboek, Koninklijke Nederlandse Akademie Wetenschappen 1961-1962:174-175.
Jaarboek, Universiteit Utrecht, 1961.
Johanna Westerdijk, een markante persoonlijkheid. 1963 (by Maria P. Lohnis, Wageningen, 95 pages)
Journal of General Microbiology 32:1-9, 1963.
Mededeling, Nederlandse Vereniging Vrouwen met Academische Opleiding 28:9-11, 1962.

Mycopathologica et Mycologia Applicata 17:359-362, 1962.
Tijding over het Katholiek Hoger Onderwijs in Nederland 4(7):15. (Mar. 1963)
Tijdschrift over Plantenziekten 67(6):549-553, 1961.
Vakblad voor Biologen 41:229-231, 1961.

Among articles describing the history of the CBS Laboratory:

Coolia 22(4):90-97, 1979.
Natura 76(8-9):205-210, 1979.

Among articles describing the history of the WCS:

Phytopathologisch Laboratorium 'Willie Commelin Scholten' 18 dec. 1894–18 dec. 1969 (by L. C. P. Kerling, Mededeling 75 of the W. C. S. Laboratory, Baarn, 1970).

Is the Elm Disease an Infectious Disease?

Johanna Westerdijk
1928. *Is de iepenziekte een infectieziekte?* Tijdschrift der Nederlandsche Heidemaatschappij 40(10):333-337.
[Journal of the Dutch Heath Society]*

[p. 333] Whether the elm will perish by the elm disease, or whether its cultivation can be continued in our country, is a question of national interest. The Heath Society [Heathland Reclamation Society] has given evidence of its great interest in this [matter] by making available for 1927 and 1928 the proceeds of the "**Prij[s]vrageninstituut of the Oranjebond van Orde**" for a renewed scientific investigation into the cause of the elm disease.

I was requested to undertake this research. Since such a study requires daily attention, I asked Dr. Christine Buisman to carry out this work.

A detailed report of this research will appear at the end of this year. But it seems desirable to me that the firm results [should] be communicated briefly in this journal already, although in a somewhat more elaborate form [p. 334] than I did at the recent Congress of the Heath Society.

To the results of Dr. Buisman, who now has provided the precise evidence beyond all doubt that "the elm disease" is an infectious disease and that the fungus *Graphium ulmi* Schwarz is the cause thereof, I want to have prefixed some observations about the nature of the elm disease. As this journal comes mainly into the hands of foresters and practically oriented people, it seems to me very desirable to set forth the newer scientific concepts on infectious diseases.

From various articles that appeared in this journal in recent years, it appears to me that the newer concepts in general have not filtered through to the practitioners. I have in mind here Mr. G. Houtzagers's two articles in the issue of 1 Sept. 1924 and the issue of 1 June 1927, and the articles of Mr. C. A. L. Smits van Burgst in the May and July issues of 1925.

For me there never has been the least doubt that the elm disease would be an infectious disease and not a physiological disease, and for the following reasons:

The first symptoms of this disease that one observes are the rapid drying out and dying back of young twigs on the tree in random parts of the crown. Sometimes the disease begins with one twig, sometimes with several at once. After [a] short or longer time the disease penetrates to other branches, which are brought to [the point of] death in a fierce way. [The] twigs and leaves look as though they are scorched. A general climatological influence, whether

*[Translated by F. W. Holmes, 8/4/1983]

this comes into effect from the soil or from the air, would much more likely affect the entire crown simultaneously; it would cause a much more general effect. One may think, for example, of the influence of drought on birches, in which leaves turn yellow and fall off early throughout the tree. In certain cases parasites can have a general effect on the tree, too. Only with the occurrence of root parasites and of fungi in the central woody structure of the trunk, can defoliation and branch death occur over the entire crown.

[p. 335] The local, very strong development on certain twigs, which only slowly spreads, is precisely what indicates the existence of a parasite in these parts.

A second evidence for an infectious disease is supplied by the manner of spread. In roadside plantings [street trees] one sees, in most cases, that the disease begins at one end and that, one after the other, the next tree in line is affected. But also if the first diseased tree lies in the middle, usually after a short time its neighbors [are the] next [to] become diseased. With general soil- or air-influences, such a development is very unlikely.

A third evidence for the parasitic nature is that the strongly [actively] growing young parts of the tree are the very ones [to be] affected.

And, to me, the fourth fact has not the least weight in the scale, to wit, that the elm disease occurs on all soils.

One finds it on sand, on clay, on more peaty spots—where elms stand dry and where elms stand wet!—and on all these spots it suddenly arises and spreads out steadily.

In 1921 we indeed had a very dry year and in that year, when one scarcely knew the disease, it seemed as though it was caused by drought. Since then we have had cold and wet years and dry and warm [years], but the elm disease proceeds unabated. That one of these conditions should have caused the elm disease is impossible, but even so this parasitic disease is under the influence of external conditions. On very dry, hot days one sees the acute dieback arise much more violently. But each parasitic disease occurs [expresses itself] to a greater or lesser degree, according to "the weather," which influences the parasite as well as the host. I need only recall to [your] mind here [that] the potato disease ([caused by a] Phytophthora), a typical parasitic disease, occurs in our wet climate almost always [every year], but nevertheless in very dry years is practically absent.

I must here, then, also object to the opinion which is prevalent among very many in the industry, and which Mr. Houtzagers also pronounces on page 180 (1927), where he opines that parasites (he assumes these [are] "Micrococcus ulmi Brussoff") [p. 336] principally affect trees that are doing badly, which, for example, suffer a blockage of the air to the roots. No, the stronger the parasitic characteristics a fungus possesses, the better it can attack very healthy, strongly growing trees. The dying off of so many very healthy, beautifully growing elms, is exactly the proof that they are affected by a deadly parasite. In the human world it happens just that way. There the very strong bodies are the very ones often toppled by infectious disease.

84

The weaker fungi, like *Nectria cinnabarina*, which only now and then behave as parasites, are found on the weakly growing [trees]; *Graphium ulmi* is not this way.

The fact, which Mr. Smits van Burgst brings up on page 144 (1925), that in many places one sees healthy trees standing between dead and half-dead, is a question of infection chances, [and] certainly does not prove that just the affected trees had a bad root condition and the intervening ones [did] not.

Anyway, the so-called healthy (outwardly healthy) trees, which stand between diseased [ones], almost always turn out to be affected if one cuts into them.

I must equally oppose an assertion in the same article (page 144) that [during] these last years the trees have had to conduct a tough fight with the natural elements. For ages, the elm has been feeling at home and [it has] developed well in our climate. It has certainly gotten through bad years many times over the centuries. But this has not been the downfall of the tree, which is so typically "at home" here that it "serves in all places."

No, at the moment it has to fight a severe battle with a parasite. It is impossible for me to be optimistic about the outcome. The Dutchman fervently wants his elm to weather the storm, and so he likes to think that it won't get that bad and that the elm disease is only a transitory disease. How was it with the oak mildew, which suddenly appeared in 1907? Has this already disappeared? Has the influenza "departed"? Then I also cannot concur with the opinion of the committee that undertook to study the question of which tree should replace the elm [Commissie in zake Vervanging van de Iep, i.e., Committee on Replacement of the Elm, five members appointed by the Minister of Interior and Agriculture on 9 April 1927, which published a 35-page report in 1928], that the elm disease may well disappear. We have no basis whatever to assume this of a severe parasitic [p. 337] disease, especially where not a single material or method for control is available.

True, the elm is propagated almost exclusively in a vegetative way. This has led to the propagation of particular types. Sowing of elm is done only to produce understock. This leads the advocates of the senility [senescence] theory to say that the elm as a species is degenerated and therefore susceptible to disease. I'd prefer to account for the affair differently.

By [their custom of] vegetative propagation, people have cultivated only particular types of elm. Should these, now, happen to be susceptible types, then with an epidemic most of the trees must succumb.

Therefore, to my mind, there is only one way to try to get rid of the elm disease, and that is to seek resistant types among seedlings. With the injection method of Dr. Buisman [i.e., inoculation by injecting fungus into the tree], it can be ascertained in [a] short time whether or not a type has a capacity to resist [has resistance]. No one can predict whether such types exist; to date we don't know [of] them.

But on the other hand, the elm culture is of such importance for our country that, in my view, people must leave nothing undone that has even the least chance.

The placing of chemicals in the ground makes no sense with this kind of disease. In the detailed report this is gone into further. Propaganda is made in this direction, since there are those with [vested] interests, who [would] like to sell chemicals. People could better save themselves the trouble.

It is not possible to eradicate the parasite of such a widely distributed disease. But also it is not advisable for that reason to let diseased trees stand for years, allowing millions of fungus spores to be spread through the air. [Bark beetle transmission of this pathogen was not yet known in 1928, but, of course, when carried on the beetles the spores *do* go "though the air."] An affected tree will die anyway, in spite of the fact that unpracticed eyes think they have seen elms recover.

People are sentimental about each tree that is cut; let them not be this way for a diseased elm: the earlier it disappears, the better.

There is little one can do against the elm disease, but let that little then be done.

Baarn, August 1928.

The Elm Disease, Report on the Research Conducted at the Behest of the Dutch Heath Society

Westerdijk, Johanna, and Christine Buisman.
1929. *De iepenziekte, rapport over het onderzoek verricht op verzoek van de Nederlandsche Heidemaatschappij.*
Uitgave Nederlandsche Heidemaatschappij te Arnhem (Holland-Drukkerij, Baarn).
[Published by the Dutch Heath Society in Arnhem (Holland Printers, Baarn): iv + 78 pages, 39 references, 15 photos, 9 drawings, 7 tables, 4 "Parts" (with 11 chapters in parts II and III).]*

[p. i]   FOREWORD

At the end of 1926 the Board of the Dutch Heath Society asked me to launch a renewed investigation into the causes of the Elm disease.

This investigation was made possible by this Board, in consultation with the Executive Commission of the "*Prijsvrageninstituut Oranjebond van Orde,*" by making available the necessary means from its funds.

Even though according to the constitution of the Institute—if the state of the finances permit it—once each 5 years a sum of at most 2,000 guilders is made available for the writing of a prize essay or for other meritorious work on the subjects of political economy, rural economy, agriculture, and [of] a general scientific nature, due to the importance of the Elm question, approximately 5,000 guilders has already been spent for this in the past 5-year period, 1924–1928, while in the time period that meanwhile has begun, 1929–1933, a sum of 1,000 guilders has already been appropriated for the same goal.

At this point it is appropriate to give a salute of honor to the memory of J. Hora Siccama van de Harkstede, Esq., through whose altruistic work all of this has become possible. After all, he was the founder and propelling force of the Oranjebond van Orde and the Essay Contest Institute called into being by this Association. At his initiative in 1923 the possessions of the Oranjebond van Orde were given to the Dutch Heath Society, which created a separate Oranjebond Van Orde Fund for the control thereof.

The Executive Commission of the Institute currently consists of Messers I. C. J. Kakebeeke, Inspector of Agriculture, Chair; Prof. Dr. W. C. Bordewijk, Professor at Groningen; and Ir. V. R. IJ. Croesen, Chair of the Netherlands Agriculture Committee, Secretary.

The Board of the 'Willie Commelin Scholten' Foundation allowed the work

---

*[This translation includes pages i–iii and 1–12: Foreword, Table of Contents, and Part I, "Earlier investigations and general considerations," by Johanna Westerdijk. Translated by F. W. Holmes, 6/1985. The rest of that publication, all by Christine Buisman, has also been translated and is available, at cost, from F. Holmes.]

to be carried out at the laboratory and on the property of the Foundation. The results of this research are set down in this treatise.

With pleasure I grasp the opportunity to thank several people for the strong help which they afforded us: Prof. A. J. Kluyver at Delft for the help with the bacteriological research; Mr. H. van Oordt, Chief Engineer of the "Rijkswaterstaat" [national agency for waterways and national highways], Middleburg; Mr. S. G. A. Doorenbos, Director of the Municipal Plantings in The Hague; Mr. G. J. Boeschoten, Supervisor of the Municipal Plantings in Leeuwarden; Mr. J. Goossen, Conservator of the Botanical Garden at Baarn; and Dr. J. Richter at Berlin-Dahlem for his help with the German text [of the summary].

—JOHANNA WESTERDIJK

[p. ii]   [List of topics in] CONTENTS*

Part I

EARLIER INVESTIGATIONS AND GENERAL CONSIDERATIONS
By Johanna Westerdijk

Part II

RESEARCH INTO THE CAUSE OF THE ELM DISEASE
By Christine Buisman

INTRODUCTION
Chapter I. Appearance of the Wood of an Elm Suffering from Elm Disease
Chapter II. Culture Experiments
Chapter III. Inoculation Experiments
   Section 1. The elm material used for the inoculation experiments
   Section 2. The *Graphium ulmi* material used for the inoculation experiments
   Section 3. Time, place, and manner of inoculation
   Section 4. Results
      a. Inoculations without prior wounding
      b. Inoculations after wounding on above-ground parts
      c. Root inoculations
      d. Inoculation experiments with 2 organisms and with extracts from diseased twigs
      e. Control experiments
   Section 5. Inoculation experiments on other tree species
Chapter IV. Discussion of Inoculation Experiments
Chapter V. Natural [Inoculations/Infections]
Chapter VI. Symptoms and Occurrence of the Elm Disease
   Section 1. External symptoms
   Section 2. Susceptibility to the disease, of trees of differing location, age, and species
   Section 3. Internal symptoms and course of the disease
   Section 4. Duration of the disease process

*[This list of the topics in the Contents, without the original page numbers, which are unnecessary to the present text, is included to show the scope of the research on DED commissioned by the Heath Society, directed by Johanna Westerdijk, and performed by Christine Buisman.]

PART III

DISEASES OF THE ELM WHICH IN ONE OR ANOTHER RESPECT RESEMBLE
THE TRUE ELM DISEASE
By Christine Buisman

PART IV.

[p. 1]
PART I
Earlier investigations and general considerations
by Johanna Westerdijk

[p. 2 was blank; p. 3] As to the research about the cause of the elm disease, I supply at the outset a very short discussion of the work that has been carried out to date on this disorder.

I am speaking here about the *scientific* investigations; among these, some have brought progress in our knowledge, others have led this [knowledge] decidedly in the wrong direction. The observations in the horticultural industry press have in this case not contributed much to our understanding, especially since there is still quibbling all the time about the cause of the disease.

It is especially the wholly erroneous scientific investigations that are spotlighted in the horticultural industry press, and they have been really dangerous to the tree-nursery community of Western Europe. I mean those investigations in which it is suggested that the soil of Western Europe is poisoned by the parasite of the elm disease. In these times of plant quarantines, only a triviality is needed for other lands to close their borders to all our trees. Reckless deductions by scientists, in this area, should be severely condemned.

In the literature discussion I have gone into only those points that in my

89

opinion are essential. How "the public" has reacted to the elm disease clearly demonstrates that "the elm" is a tender spot in the soul of the Dutch folk. Even though I shall make a few observations about this reaction, I don't go into the innumerable popular articles and lay observations, which have literally "harassed" a correct understanding of the disease. For that matter, I do not delude myself that this chatter will come to an end after this report. The big [national] press hardly accepts lay opinions about human diseases any more: in the area of plant diseases they still swallow every indigestible lump.

At the end of this literature review, I give a short summary of what the renewed research in the Willie Commelin Scholten Phytopathology Laboratory has yielded. Finally, I give some general considerations about "attempts" to stop the disease.

Literature Review

In Mededeelingen van den Phytopathologischen Dienst [Communications of the Phytopathological Service] No. 18 (1921), D. Spierenburg discusses for the first time the occurrence of the elm disease, which has been observed in our country since 1919. She describes brown rings and streaks in the wood, and isolates a number of fungi, including a *Graphium* and a *Cephalosporium*; she concludes, however, with no single disease cause, and does not know whether the disease is of inorganic or infectious nature. Typical, however, is [p. 4] that she provides as "opinion from the industry" that they think they are dealing with an infectious disease, which begins in the top and the tips of the higher branches.

In Communication No. 24 of the Phytopathological Service 1922 she continues her studies on "an unknown disease in the elms." She describes different disease expressions, gives lists of fungi and bacteria which she isolates from the diseased wood and from the bark, and comes to the following results: She can demonstrate no fungus in the discolored wood; can mostly culture the fungi *Cephalosporium-Graphium* from the wood; sometimes besides these, bacteria. After inoculation with these organisms she sees only dark specks arise in the wood, and still wishes to make no pronouncement about the cause. Further, she points out already an acute and a chronic stage of the disease and its occurrence on all soils.

Although this treatment gives a clear insight so far as the disease pattern is concerned, as far as the cultural part is concerned this investigation is hardly refined: because, with pure culture, it is only *Graphium* that always comes out of the wood. Also it can be demonstrated without difficulty, by single spore cultures, that the little *Cephalosporium* heads and the *Graphium* coremia belong to one and the same species. And reisolating bacteria from trees inoculated with fungi is always an indication, in this connection, of not very clean work.

A few weeks after this communication, the dissertation of M. B. Schwarz appeared: "The Twig Dying of Elms, Weeping Willows and Peach Trees," based on work in the "Willie Commelin Scholten" Phytopathological Laboratory.

The essence of this work is that the mycological work is, in contrast, very precise; with pure culture, only *Graphium* and no other organism was isolated from the diseased wood, while it was proved that the form with little heads (*Cephalosporium*) is one of the forms of *Graphium*: this fungus was now identified as *Graphium ulmi* nov. spec.

Schwarz finds hyphae in places in the wood vessels. Further she made inoculation experiments (which were made mainly in the autumn) in which she got brown wood discoloration over [a distance of] up to 30 cm vertically in a branch. She did not get acute dieback symptoms, which we can now understand, because inoculations were done in the wrong season. Schwarz thought nonetheless that she must infer from this (and I thought this with her) that *Graphium ulmi* is the cause of the elm disease. About [this] conclusion her work is continually attacked in a very vehement way in all sorts of newspaper articles. The writers of these essays apparently do not realize in how many cases of well-known diseases inoculation experiments have *never* been done. Her idea that the infection should occur through the leaves, has not materialized [has proved false]. Anyway, the acute symptoms still had to be evoked.[1]

[p. 5] However, before this happened, there arrived the period in the literature that we can label as the "bacterium-period"!

Because in 1924 Brussoff at Aachen [Germany] published an article in which he very clearly described the anatomical characteristics of the elm disease, but in which he came to the conclusion that a bacterium "Micrococcus ulmi" nov. sp. is the cause of the disease. He thought this on the basis of cultures and of the round globules that he observed in the vessels.

His description of the diseased wood is so typically that of *Graphium*-diseased wood that one need not doubt that he has had it in his hand. He does not further consider, however, the "yeast cells" in his cultures (possibly the little-head form of *Graphium*).

His inoculation experiments, to which he ascribes positive results, prove nothing: because he has inoculated large trees in the trunk with bacteria and then seen that in the summer some tops died back. If a tree so treated is not sawed into slices after the removal, by which the connection between the point of inoculation and the diseased tops is demonstrated, the experiments are without any value or conviction. It does not appear from Brussoff's

[1]In the "Communications of the German Dendrological Society 1922," Valkenier Suringar [Professor of Dendrology at the Agricultural University in Wageningen] gives a brief discussion of the results of the investigations of Spierenburg and Schwarz.

In this he calls *Graphium ulmi* nov. spec a "nomen dubium." In general, one can say that dendrologists never involve themselves with the describing of fungi and [p. 5, bottom] know nothing about it. This meaningless criticism of this fungus, which was described according to modern methods, I surely need not refute here.

Later this same author once more has obliged us with a sort of criticism of the elm disease researches (Yearbook of the Dutch Dendrological Society, 1927) in which *Graphium ulmi* must suffer again. Now it is even called "Graphium doctorale."

Allegations like this are of such [an] inferior alloy that one actually does not understand why the Dendrological Society lets them be printed. Such a thing has nothing to do with scientific criticism.

publication that this has been done. Moreover, his judging of the trees to be infected is very primitive, in that he only sees whether the drilled hole has brown wood.

About the cultural part—the isolation experiments—in which he describes *Micrococcus ulmi*, I need to say nothing at all further, since Stapp (see below) has shown that no *Micrococcus ulmi* exists.

In 1926, Brussoff published some very alarming articles, in which he asserted that the elm bacteria did not remain limited to the elms, but attacked all sorts of trees. Maple species, linden, beeches, poplars all show, in his opinion, disturbing dieback symptoms, which are the consequence of a bacterial infection in the roots. However, it is striking that his descriptions of the wood infections seen with the unaided eye, hardly remind [one of] the *Graphium* streaks. Commonly he talks of flecks and dots and of red discolorations. It is very deplorable that these communications, which rest on extremely doubtful experiments, have found their way into the horticultural press everywhere. Also, by a report in a widely read periodical like "Umschau" [a popular German general magazine that reviewed agricultural subjects], everyone was terrified that the roots of almost all [of] our tree species were contaminated with a bacterium [*Micrococcus ulmi*] with which the ground was literally drenched. Also, in the communications of the [German] Dendrological Society, he gives a portrait of a diseased linden root that very strongly reminds one of an attack by *Verticillium*.

[p. 6] When the alarm bell was so urgently sounded, the [German] National Biological Institute at Berlin-Dahlem decided to have an investigation into the elm disease started, both by its bacteriologist Dr. Stapp and by its mycologist Dr. Wollenweber.

Before we discuss the results of these investigators, we still have to mention the work of Countess Von Linden and Lydia Zenneck at Bonn, that appeared in 1927. They isolate *Graphium ulmi* Schwarz from all diseased branches, and conduct experiments by putting branches into an extract from cultures of this fungus, whereupon they observe wilting. S. Broekhuizen has harked back to these experiments in a paper on the causes of tylosis formation in the elm.[1]

In the same year Brussoff made a critique of the investigations of Linden and Zenneck, while these latter responded to it. For the course of the research on the elm disease, these articles are of little importance and therefore they can remain outside [this] discussion.

In March 1927 we began a renewed investigation on the elm disease. In the same year Wollenweber and Stapp at the [German] National Biological Institute began their research. In October 1927 Wollenweber gave a brief survey of his experiments in the "Nachrichtenblatt" [Report Sheet] of the German Plant Protection Service. Wollenweber confirms the assertion of Schwarz that fungal hyphae can be found in the wood, and that these belong to the fungus

[1]S. Broekhuizen. Wound reaction of wood. Leiden. 1929.

Graphium ulmi Schwarz. Besides seeing the fungus penetrate the wood after inoculation, he sees wilting and withering arise 5 weeks after the inoculation of 1- [and] 3-year-old seedlings of *U. montana* [*Ulmus glabra* Huds.]; the spread of this fungus occurred upward and downward and gave the typical appearance in the wood. The inoculations took place in July.

By this, exact evidence is provided that, besides the discoloration of the wood, the withering also is brought about by *Graphium ulmi*.

An extensive treatment by Wollenweber appeared in 1928. Together with Stapp, he gives his conclusions. Wollenweber again describes numerous successful summer inoculations with *Graphium ulmi*, mainly on 1- to 3-year-old *Ulmus montana* seedlings. This researcher always makes use of a T-cut, into which he puts the fungus from a pure culture. This is a different method from that with which Buisman (see Part II) had success. The infections extended themselves in the course of the summer from the bases of the small trees to a height of 1 meter. These experiments are convincing.

Experiments on larger trees, which were started in the autumn, tell less, since the tree was not felled and sawed into pieces, but was judged exclusively by external symptoms. In this way it is not possible to be completely certain of the success of inoculations. As to this point, one cannot be careful enough in one's judgment about larger trees.

Wollenweber totally agrees with the diagnosis of Schwarz, and likewise considers the elm-*Graphium* as having been undescribed prior to 1922. Further, it is also very important in Wollenweber's research that he finds *Graphium* spores in the galleries of beetles.

[p. 7] In the same publication, Stapp explains how no Micrococcus can ever be found in the authentic cultures of Brussoff, and how he has not gotten even one infection with any isolate of Brussoff's. One can safely say that "Micrococcus ulmi" is out of the picture. Brussoff's defense against this research [by Stapp] is meaningless. Moreover, Stapp can never isolate bacteria from "elm-diseased" elms.

In recent months a few more communications about the elm disease have come, and these from Malcolm Wilson and Dr. Mary Wilson. Both culture *Graphium ulmi* from the elm-diseased branches [twigs] and believe that in England, too, the elm disease is caused by *Graphium ulmi*.

Further, they have found spores of this fungus on dead stumps.

In 1927 there appeared also a treatment by the well-known mycologist Ph. Biourge in Leuven [Belgium]: "La maladie des ormes" ["The elm disease"]. First he found a case of *Nectria*, which he then considered as *Nectria cinnabarina*, on a branch of a couple of large, dying elms. Some time later he sees the paper by Eriksson (cf. No. 13), who found *Nectria* on twigs which had been killed by *Exosporium ulmi* the previous year. Then he gets [a chance] to see the first publications by Spierenburg, which he discusses at length.

From diseased branches, then, Biourge isolates *Nectria* with its *Tubercularia* conidia. He believes [that] he sees this *Nectria* convert, in culture, into *Graphium*. Thus he concludes that *Graphium* belongs in the cycle of *Nectria*.

93

He even goes much further, and will implicate in this cycle not only *Exosporium* but all kinds of other spore forms that one finds on elm branches, e.g., *Phyllosticta ulmi* Westendorp, *Libertella ulmi-suberosae* Oudemans, *Gloeosporium inconspicuum* Cavara. It is not clear to us whether these considerations are based only on raw [newly isolated] cultures. In any case he has never cultured and subcultured separately *any* of the many spore forms that he found. Such a way of working is out of date, and one can draw not a single definitive conclusion from this. Therefore it is superfluous to go into the interrelationships of all the forms that he names, especially since his deductions are extremely obscurely stated.

Likewise, his considerations as to which *Nectria* he really is dealing with are very confused. He believes that they belong to *Tubercularia nigricans* (=*Tremella nigricans* Bulliard) and thus regards *Nectria nigricans* (Bull.) Biourge as cause of the elm disease. It is not possible for us to discover any thread in these confused speculations. He gets his data principally from the era antedating pure cultures. He has not experimented himself.

In 1928 Dr. Buisman and I published a preliminary little report on the new elm disease investigations (1927, 1928), from which it appeared that here numerous artificial inoculations with *Graphium ulmi* could evoke the elm disease in closely inspected trees 3 to 6 years of age.

Moreover, Dr. Buisman found that black streaks could occur in elm branches which showed no outward disease symptoms. These comprise vessels which are filled with bacteria. It is a vessel infection that is not [p. 8] of profound influence, but which can easily be confused with *Graphium* vessels upon cutting through the twigs. These bacteria have nothing to do with "Micrococcus ulmi:" they are rod-shaped bacteria. They do not bring about disease symptoms. Further, Dr. Buisman investigates some other diseases of elms that could be confused with Graphium.

In the report submitted here these investigations are extensively expounded.

Conclusions

The conclusions that one can draw from this report, I would summarize as follows:

Graphium ulmi Schwarz is the cause of the elm disease, because by injection with this fungus all typical symptoms of the elm disease have successfully been brought about. *Graphium ulmi* is a typical vascular parasite, which spreads mainly longitudinally in the branches and trunk, and very little horizontally. Natural infection occurs through the branches [twigs], progresses through the trunk and can finally affect the roots. Despite success in artificially infecting roots, this probably does not happen in nature, certainly not as a rule. There is no question of "poisoning" of the soil with this parasite. Infection does not take place through the natural openings of the plant but, as one must assume, through small wounds in twigs. Artificial inoculations on unwounded branches [twigs], leaves or flowers do not succeed; nor do inoculations on pruning wounds. The parasite can grow out into younger branches [twigs]

94

during a year following the infection. More young branches die by infection from the outside in the one year than in another year. A tree can be diseased, i.e., have vessel discolorations, for years before one sees external symptoms. The young infections are not always visible at once. The higher the summer temperature, the better [more] visible they are. The acute death is greatest in times of heat-waves. The spreading of the old infections within the tree, by which vessels are put out of action over long distances, is usually visible [at] the beginning of June. Infections take place from June to the end of August.

A diseased tree does not recover; whether the disease becomes "dormant," or whether it kills the tree in a few years, depends on the age of the tree and on the summer temperature.

All phenomena of spread indicate that the spores of the fungus spread through the air with the wind. We have found no spores on dead branches or trunks but the research into this must be laid out on a very broad scale. In England, however, they have found spores on dead stumps.

Branches with "discolored vessels" by no means always suffer from the elm disease. The different discolorations, which are based on different causes, however are perfectly distinguishable by [means of] accurate study.

There is no question of our knowing "everything about the elm disease." However, this is not true of any plant disease whatever: but the better our methods of [p. 9] research become, the more particulars about the disease we shall get to know.

General Considerations

Spierenburg [has] already pointed out, in her publications, that everyone is marketing wisdom about the elm disease and that everyone presumes to offer criticism[s] of the most limited statement by experts regarding this matter.

After years [of experience], I should like to go even a trace farther. I should like to say that the elm disease gives inducement to human psychosis. Newspapers include [everything] without selectivity, in order to "let all opinions be expressed." That there exist "opinions" and opinions they have more or less forgotten. The Dutch fear for their elms, would rather not hear that the elm disease is an infectious disease, and are delighted if a critical remark is made about inoculation experiments, whereby they then can fall asleep again, with the idea that it "is a temporary disturbance of the elm." Lay people give lectures and jabber: an artificial inoculation experiment has nothing to do with nature, etc., etc. I have already said earlier: the course of the disease is so typically that of an infectious disease that I still would not doubt it even if no parasite had been found yet.

One can also lay at the door of the "psychosis" [blame for the fact] that for a while people let themselves be talked by a manufacturer into using manganese sulfate, which was pushed as a remedy because it cured the (peat-district) "soil disease" of the oat. There is not a single reason to assume that this has any effect on a vascular infection.

Cure and Control

After it is [has been] determined with assurance what the cause of the elm disease is and now that a large part of the life of the parasite is known, the question arises: is the disease controllable?

When we watch for some years, it is evident that a diseased tree does not cure itself. In the spring, when the leaves have just come out, it sometimes seems as though the disease is conquered, but in the course of the summer it turns out indeed that this has not been the case. If we let a diseased tree stand, then it gradually succumbs, be it within a few or after many years.

A control measure that is applied in the industry is the so-called "candalabring" [pollarding, cutting all branches drastically back]. Theoretically, of course, if only a few infections are as yet present in a tree, one can cut these out. By this strong cutting-back, therefore, they have indeed sometimes cleaned up a tree. But of course one never knows whether all infections are cut away, and ordinarily such a tree nevertheless remains exposed to infection [inoculation] from the vicinity, and sooner or later it is again diseased. So this method is seldom sufficient.

"They" ask for a material with which the roots [i.e., the soil] can be irrigated, which is taken up [by the roots], kills the parasite and moreover does not affect the tree tissues. That such materials should exist is very unlikely. With this, one must consider first of all that a root chooses and does not take up everything off-hand. [p. 10] If the materials in question [should] be poisonous to living cells, then at the outset they will [would] kill the root hairs and then the tree can [could] take up nothing more at all. "They" also contrive medicines with which a tree must be injected and so stimulated that the disease is conquered. With the current state of our knowledge these measures offer extremely little chance of success. I [have] already mentioned, above, the manganese sulfate [instance].

Supposedly favorable results in the field, by manuring and injecting with chemicals, are—insofar as I know—attributable to impure experiments. There was never any mention of comparable "check trees." Many a tree looks better in the spring than in the previous summer since the young twigs leaf out in a healthy way. Here too, the previously reported lay and half-lay talk has done a lot of harm.

In modern medicine, immunizations and vaccinations play a large role. The concept "serum therapy" cannot, however, be carried per se over into phytopathology. A tree does not have a circulating sap-stream that can be compared with the blood circulation of people; the sap of a plant and blood are not comparable. If one injects a solution into the trunk of a tree, it is out of the question that the material should reach all vessels of all branches.

As a direct control measure, then, there would remain the destruction of the spores of the fungus by chemicals. From the few discoveries that people have made, it appears that these form in dead bark or on stumps of dead elms.

For a tree with a large crown, as an elm has, it is just as infeasible to destroy the spores on dead trees at once, as [it is] to spray healthy crowns as a protective measure. The infection comes while the crowns are foliated and then one can never contact all twigs (June - July - August). It is still possible that they will discover other hiding places of the spores or that they will find the winter spores (ascospores) of the fungus but this has not yet happened.

As an indirect control measure one could suggest the destruction of diseased individuals in order to limit the chance of infection [inoculation] of the healthy trees.

This kind of control can, by its nature, be applied to the matter only in districts where the disease still occurs only extremely sporadically. For our country such a measure would make no sense, since the disease is much too general [in distribution] for this [measure], and the air must be heavily laden with spores. In special cases (for example with long, elm-lined streets where a single individual is affected) one can perhaps stem the disease by cutting down the first diseased individuals, since then the neighbors become diseased less quickly. If the whole row is lightly affected, however, then one can let the trees stand until the street becomes ugly, without having to be anxious that the chances of an infection might increase considerably for the whole district, since these [chances] are very great over our whole country anyway.

It is entirely another thing, naturally, when we come into districts where [p. 11] the disease is just beginning to arise and so occurs sporadically and often is spreading from a single focus.

Then, at least, one can make attempts to halt the disease by eradicating such foci and also by cutting down neighboring elms, which may already be infected, albeit lightly and not visibly.

In England, where they first observed the disease several years ago, they thought they would have to decide on [to use] this method. Upon a systematic inspection of the elms, the disease turned out already to be much more widespread than they originally had thought so that it turned out, for all practical purposes, to be infeasible to cut down and burn all affected individuals.

In a country like Denmark, for example, where *Graphium* is not yet recorded, eradication of diseased trees at the first infestation would provide a small chance of limiting the disease.

It has appeared to us that diseased mother trees ["stools," or source trees for production of layered cuttings] do occur in nurseries in The Netherlands. They certainly form a source of infection, even though young trees are decidedly less susceptible than older ones. On the other hand, in nurseries trees are regularly pruned back and transplanted, by which means young infections are removed directly.

In our opinion, there is only one path which one can tread in order to try to get healthy elms again. The elm which is planted along our dikes is

97

a tree that is layered, a tree which continually is asexually propagated. This entails that a uniform type is planted out. With "seed trees" like oaks and conifers the type is more divergent. Our type—our Dutch dike elm [*Ulmus hollandica* 'Belgica'], and also the monumental type [*Ulmus carpinifolia* 'Sarniensis'], which they like to plant in cities—[are], as we know, extremely susceptible to the disease. No one knows whether other types do not exist which are much less susceptible to disease.

The elm disease is a typical vascular disease. There are many vascular diseases well known among herbaceous crops (e.g., [those] caused by *Fusarium* species). In the case of exactly these diseases, so-called resistant races are known. With the vascular-wilt diseases of cotton, tomato, melon, flax, [and] cabbage, numerous types which are barely susceptible and even wholly immune have been selected in the United States. What the characteristics are that bring about this resistance is not known. Yet the best examples are known in exactly this category of diseases.

It therefore is not at all unthinkable that such types are also to be found in the case of the elm. The experiments of Dr. Buisman with inoculation of seedlings point in this direction. We know, too, that there are horticultural types (e g., the [*Ulmus*] *montana fastigiata*) [syn. *Ulmus glabra* 'Exoniensis'] which are less susceptible than the dike elm.

Since Dr. Buisman worked out a method by which in the summer after 10 days one can already see whether [or not] the elm contracts the disease, it will be possible in the future to judge susceptibility of elms quickly. I think, however, that these types must be tested for more than one year in succession, in order to assure that under no circumstances will they become diseased. Such types will have to be multiplied and repeatedly retested for resistance. First of all a suitable dike tree must be sought; after that [p. 12] one could test numerous species and varieties of *Ulmus* that are grown in parks and gardens.

Our knowledge of the many varieties is very limited. They speak, too, of mongrels [crossbreeds, hybrids] as though they had seen them arise: more taxonomic knowledge of "elms" ought to be able to help us with this.

The elm types must be collected from our country and from other countries. The inoculating of these trees must occur judiciously in an experimental plantation with sufficient individuals to discard trees repeatedly.

Here, the way that can lead to this goal is long, much longer than in the case of herbaceous plants, especially since the elm is more susceptible as a twenty-year-old than as a two-year-old tree. Still this is the only thing to be done. For The Netherlands, where the elm is the most important tree for our unique low lands, the maintenance of this tree species must be sought. One will have to find a type that need not suffer in so great an extent as those now planted out.

The search for another "Dutch elm" is the only thing that can be undertaken against the elm disease and it would be irresponsible not to try this. It is a Jan Salie policy [spineless]* to "believe" that the disease will disappear by itself. [It is] fortunate that our medical people have never cultivated this belief with regard to our diseases. Let us also not do it for our elms.

*[This is a reference to Dutch literature: in a book by Potgieter, *Jan Jantje and His Youngest Son* (1842), Jan Salie was a character who typified lack of energy.]

Dr. Chistina Buisman, 22 March 1927. This day was both her 27th birthday and the day when she defended her dissertation and received her doctor's degree.

CHAPTER 5

Christine Johanna Buisman

1900-1935

Christine Johanna Buisman, called "Stien" by her friends, was born March 22, 1900, in Leeuwarden, capital of Friesland Province, The Netherlands. She had one sister, Sophie Alida Catharina Buisman. Their mother, Hillegonda Cornelia Blok Wijbrandi, was active in politics and social work and was a member of the city council. Their father, Roelof Buisman, was a merchant.

In 1919 Christine Buisman entered the University of Amsterdam, where she completed her *doctoraal* in 1925 in biology.*

From February 1, 1925, until March, 1927, Buisman held an assistantship in the Centraalbureau voor Schimmelcultures (the national fungus culture collection) in Baarn. Here she finished her doctor's degree in plant pathology under Professor Johanna Westerdijk. Her doctoral dissertation on root-rotting caused by lower fungi (phytomycetes)—treating, in particular, the rotting of calla-lily roots—was the only publication in her entire career that treated any disease other than the one we now know as Dutch elm disease (DED) and other elm diseases that may be confused with DED.

By 1929, with financial support provided by the Dutch Heath Society, Buisman conducted a two-year study, which was amazing for its large number of experiments, that showed that Schwarz had been right: *Graphium ulmi* is indeed the true cause of DED. Buisman then found the ascospore stage— first in the laboratory, and later in nature—and named it *Ceratostomella ulmi*.

The career of this most promising and talented young scientist was like

---

*[Both Americans and the Dutch often confuse the English adjective "doctoral," which means "of a doctor's degree," with the Dutch noun "doctoraal" (formerly an adjective in "doctoraal examen"), which is somewhat comparable to a "Master's degree" in the United States. Dutch people sometimes say that a doctoraal, which can take 5 or 6 or more years, is "heavier" than the Master's degree in United States. On the other hand, the American candidate–in science at least–is expected to spend full time working on the master's thesis research and related study for the degree, whereas the Dutch candidate often completes the degree while also employed more-or-less full time in university teaching. There are differences over time and among the different universities in each country, as well as between the countries. A further confusion has arisen in the two countries, about the Dutch abbreviation "drs." versus the American usage "Drs." The Dutch use "drs." (and, for certain aspects of engineering and science, "ir.") as the title of someone who has this doctoraal degree, in contrast to "dr." as the title for someone who has a doctor's degree, whereas Americans use "Drs." as a title written before a sequence of the names of several persons, each of whom has a doctor's degree.]

a shooting star: blazingly bright but tragically brief. In her eight-year professional career, she published 32 papers, in five languages, and conducted studies on DED not only in The Netherlands, but also in a number of other countries. She did research in Germany (with Wollenweber, Richter, and Teschner), in the United States for a year beginning September 1929 (at Radcliffe/Harvard with Weston, and at the Arnold Arboretum and the Farlow Herbarium with Faull), as well as in Italy (in cooperation with Ansaloni). Her area of study in the United States was elm cankers caused by three species of the fungus *Botryodiplodia*. She also taught a course in mycology in Coimbra, Portugal, where she calmly continued lecturing when all electricity stopped as the result of a bombing of the electric station.

In 1930 Buisman identified the first finds of the DED fungus in the American Midwest. Some Dutch records (Vakblad voor Biologen No. 8, April 1936, "In Memoriam . . ." by Johanna Westerdijk) say Illinois, but evidently Westerdijk was less aware of the different U.S. states than Buisman was. The Illinois Natural History Survey has no record of any 1930 case in that state. And Buisman's *own* report ("De Iepenziekte in America," Nederlands Bosbouw Tijdschrift 7(11):439–440, 1934) says that the first U.S. DED (in 1930) was in Ohio: four trees in Cleveland and one in Cincinnati. Certainly 1930 is generally recognized in the United States as the date of our first DED cases, and Ohio is the place. Thus Buisman probably handled the first find of DED in our country. She adds that there were four more cases in Ohio in 1931, none in 1932, and a single case in 1933, whereas New York's first find (in 1932) rapidly exploded to more than 3,800 cases in New York and New Jersey by August 1934.

Collection of basic plant material for elm selection, from different sources, had started in The Netherlands as early as 1928. After Buisman returned from the United States, she took over this program (in October 1930) at Westerdijk's request and began selecting among these elms for resistance to DED. She also laid plans for, and gathered and tested preliminary selections for, an elm breeding program that she was fated never to carry out—a program that was implemented from 1935 to 1954 by J. C. Went and which is still continuing (Common Market-wide) under H. M. Heybroek, 60 years later!

Buisman's character combined an alert, clear-thinking intelligence, scientific ability, and an understanding of the practical world. From her youth she loved living nature. She was a member of the administrative boards of the Dutch Phytopathological Society and the Dutch Botanical Society. And in her last few years she began also to take an active part in the Dutch women's movement.

In 1935 Buisman entered a hospital in Amsterdam for surgical removal of a cancer. Whether because of the cancer, iatrogenesis, exhaustion from her work, or some combination of these, she died there on March 27, 1936, at the early age of 36 years and 5 days.

Published memorials to Buisman are found in:

Tijdschrift over Plantenziekten 42(7):175-178, 1936 (by Johanna Westerdijk).
Comite Inzake Bestudeering en Bestrijding van de Iepenziekte, Mededeeling 22:1-2, 1936.
Der Biologen 5(6):222, 1936.
Vakblad voor Biologen 17(8):141-143, April 1936 (by Johanna Westerdijk).

The Cause of the Elm Disease

Buisman, C. J.
1928. *De oorzaak van de iepenziekte.* Tijdschrift der Nederlandsche Heidemaatschappij 40(10) 338-344.
[Journal of the Dutch Heath Society]*

[p. 338] In the beginning of 1927, when I began my research, mainly the following opinions of the scientific investigators of the elm disease confronted one another:

1. the disease is caused by *Graphium ulmi* ([opinion of] Schwarz);
2. by *Micrococcus ulmi* ([opinion of] Brussoff);
3. by climatological circumstances ([opinion of] Pape).

Spierenburg does find *Graphium ulmi* in diseased elm twigs, but does not yet wish to regard this as the cause of the disease.

In 1927 and 1928 the following additional opinions were put forward:

The disease is caused by *Graphium ulmi* ([opinion of] Linden, Wollenweber);

–by *Nectria cinnabarina* ([opinion of] Biourge):

–not by *Micrococcus ulmi* ([opinion of] Stapp).

The question now arises: how is it possible that such variable judgements are made about one and the same phenomenon?

This has the following causes:

1. Success had not yet been obtained in 1927 in evoking an acute twig dieback after artificial inoculation, be it with *Graphium* or with other organisms. Only Brussoff had then asserted that after artificial inoculation with *Micrococcus ulmi* the trees became sick, but his experiments are very debatable and thus not convincing. During 1927 and 1928 Wollenweber reported positive results with *Ulmus montana* (mountain elm) [Scots elm, syn. *U. glabra* Hudson].

2. Various investigators could not show microscopically, in the sick wood, any organism that could cause the disease.

3. We have come to the conclusion through our investigation that in the elms we are dealing with *several diseases, which in part show the same symptoms as [those that] we encounter in the true elm disease and which make the matter very complicated.*

I shall delve further into all these points.

[p. 339] It should be stated at the outset that this research was set up completely anew, without placing in the foreground the idea of a definite fungus or bacterium.

*[Translated by F. W. Holmes, 12/20/1962]

Therefore I first had to find out, anew, what organisms could be grown from twigs affected by the elm disease, in order to carry out artificial infection tests with them. Consequently, as soon as the elm disease appeared again, approximately mid-June of 1927, I took pieces of diseased twigs, sterilized them externally, peeled them and laid them out in dishes in which I had poured various nutrient media. In this way I always isolated *Graphium ulmi* Schwarz on all the media used. Therefore extensive infection tests with this fungus had to be undertaken.

It was necessary that I have for my tests absolutely healthy elm material. Therefore it was out of the question to work with older trees, since most individuals are affected. Moreover, it is impossible to check a large crown entirely, for the typical wood discoloration in the twigs, let alone for that in the trunk. I used, therefore, principally 3- to 5-year layers [rooted cuttings] [that were] of various varieties and grown on various soils, both typical *campestris* [*U.* × *hollandica* 'Belgica'] and the variety *monumentalis* [syn. *U. carpinifolia* 'Sarniensis'].

I proceeded with my inoculations in various ways. Sometimes a flat cut was made in the twig, into which mycelium and spores of *Graphium ulmi* were then put. Another time I inserted spores by means of a needle prick between the leaf and the young bud. Also all the leaves of a twig were streaked on the lower surface with spores of *Graphium*.

In the beginning all inoculations led to practically nothing. At the most there was in the wood a dark discoloration from the spot inoculated with *Graphium*, traceable a short distance below; but there was no question of an extension of any importance, let alone of an acute death of inoculated twigs.

As soon as I employed another method, however, I obtained positive results. Namely, I put spores in water and put that [suspension] into a hypodermic syringe; with this I punctured a twig on all sides in various places. When, after some time, [p. 340] I cut off such an infected twig and peeled it, I could see how long, dark-colored streaks extended from the point of inoculation both proximally and distally. In the meantime it had already become August 1927. An acute drying of twigs no longer took place following artificial inoculations in August. The leaves of several twigs thus treated did change color in the autumn before the other leaves.

In the spring of 1928 the inoculations were, of course, continued. In the first months they had no success but in June and July I very clearly obtained positive results. I then again inoculated several trees by introducing *Graphium* into the twigs with a hypodermic syringe. In a few of the twigs I had already put *Graphium* the previous September, too, but there was nothing to be seen of it externally; they appeared completely healthy.

2 In a very large number of the twigs thus infected, some time after the inoculation (9 to 12 days) the leaves began to wilt and to dry up. This happened both with twigs that I had inoculated only this year and with twigs into which

I already had put *Graphium* in September. The progress of the disease could be followed very clearly. First the tips of the infected twigs began to dry out; gradually the withering proceeded downward until the noninoculated twigs above or below also showed the disease symptoms.

Thus a syndrome was obtained that completely agreed with the known picture of the elm disease.

Apparently the circumstances at the beginning of June 1928 were very favorable for the extension of the infection. Although the inoculations from the end of June to mid-July also gave a positive result, the disease did not then progress so quickly.

Graphium was very easily recovered from the twigs thus made sick.

None of the trees not inoculated with *Graphium*, which stood everywhere between the treated trees, show any occurrence of the disease. The twigs that I injected with water or with a nutrient medium without *Graphium* spores, show no harmful consequences from the slight wounding; only brown discoloration in the wood, a single cm. long, [p. 341] occurs at the point of the inoculation. Also the twigs which I had injected with another organism (see below) are still all completely healthy; no trace of any wilting is to be observed on them. We may therefore be confident that the drying and the death of the inoculated twigs is a consequence of the *artificial inoculation with Graphium ulmi*.

Inoculation succeeded very well also by injection of the roots with spores of *Graphium*. The long, brown infection streaks could be found within two weeks nearly to the tops of the trees.

In nature we never observed infection of the roots. We shall write a communication later about the way [route] which germinating spores actually take, since the experiments about this are not as yet completed.

We thus have been able to evoke the disease with all its typical symptoms within a very short time after artificial inoculations with Graphium ulmi. This shows that Graphium ulmi is the cause of the elm disease.

Now, as for the fact that several investigators found no organism in the diseased wood, I should like to make the following observation. After long search under the microscope it is possible here and there to demonstrate the extremely fine mycelial filaments of *Graphium ulmi*, but frequently this is very difficult. One must not imagine, then, that *Graphium* plugs the vessels with a lot of mycelium and thereby prevents water transport. Rather, *Graphium* exerts a poisonous effect in reaction to which the elm makes numerous tyloses; it thus cuts off its own water supply. These brown-colored tyloses make it difficult to see the mycelial strands.

As far as the diseases which, in certain circumstances, could be confused with the elm disease, I would name those caused by: 1. *Nectria cinnabarina*, 2. *Verticillium dahliae*, 3. *Phomopsis* sp., and 4. *a yet undescribed bacterium*.

1. Although I myself have done no research on the *Nectria* disease, I shall nevertheless mention it in connection with the publication by Biourge. Indeed,

Nectria cinnabarina, primarily with horticultural varieties of the elm, can evoke a syndrome [p. 342] that reminds one of the true elm disease. The top of the tree is affected first; the uppermost leaves wither; gradually the disease proceeds to the lower branches. There is, however, a sharper demarcation between the diseased and the healthy branches than with the true elm disease. The wood of such a diseased tree also does not show the familiar brown streaks, but dries out gradually and uniformly [without pattern]. In late summer one always finds very clearly on the branches the red sporophores of this *Nectria* species; one never finds them on *Graphium* branches.

2. Dr. J. H. van der Meer has reported earlier on the occurrence of *Verticillium* in elms. I encountered this disease only in seedlings which grew on land where *Verticillium* occurred extensively in the soil, to wit, old potato land. Just like *Graphium, Verticillium* gives a brown discoloration to the wood and also causes death of twigs. In many cases the discoloration caused by *Verticillium*, however, is somewhat watery and spreads more through the whole wood; it does not, then, form sharp streaks. Nevertheless one must sometimes wait to see which fungus one cultivates from an affected twig before being able to say with assurance whether the illness was caused by *Graphium* or by *Verticillium*.

Since *Verticillium*, however, occurs only in young, layered plants [grown from cuttings] and *Graphium* mainly in older trees, one generally can judge from the beginning with which disease one is dealing.

3. A phenomenon that seems to appear chiefly on shaded branches [of *Ulmus hollandica* 'Belgica'] is the occurrence of wilted and dried-out tips. The leaves then wilt and become brown on a more or less large portion of the branch. One could take this disease for a beginning stage of the elm disease, especially where it appears abundantly, as at the Old Scheveningen Road in The Hague. If one examines such a withered top closely, then it appears either that a little canker occurs below, or that the branch has for a short distance a wholly dead bark. Many fungus hyphae run in this bark [and] sometimes also in the wood or even in the pith that lies beneath. If after surface-sterilization, one lays out bits of such a twig on a nutrient medium, then a *Phomopsis* species, a [p. 343] fungus in a genus that is still very incompletely known, grows out. *This phenomenon has nothing to do with the real elm disease.*

4. When seeking for young stages of the elm disease, I got hold of some young twigs which displayed dark stripes in the wood. These stripes extended from the leaf scar some way downward and then suddenly stopped. When some sections of such a dark stripe were laid under the microscope, it appeared that *the vessels were plugged by a great mass of bacteria.* These were not related to the bacteria described by Brussoff—they were rods, and Brussoff mentions only round bacteria. They were easily isolated when a twig was surface-sterilized, peeled and split in two.

Although I supposed at first that I was dealing with a unique case, it soon

108

appeared that I could cut almost no other elm without encountering these bacterial discolorations in the wood. It is quite incomprehensible to me that none of the [other] elm disease investigators has yet run across them.

Externally, this bacterial disorder of trees is scarcely visible, although their foliage perhaps is somewhat sparser than [with] healthy trees. *I have never encountered acutely dying twigs resulting from the bacterial affliction.*

The dark streaks in the wood caused by bacteria are distinguishable from the *Graphium* streaks in several respects:

1. The *Graphium* streaks are not continuous but repeatedly interrupted; the bacterial streaks are always continuous.

2. The *Graphium* streaks are mostly dark brown; the bacterial streaks are mostly black.

3. The *Graphium* streaks are not sharply delineated (the discoloration spreads [peters out], as it were, into the adjoining wood); the bacterial streaks are sharply discrete.

4. The *Graphium* streaks are difficult to follow; the bacterial streaks [are] mostly easily [followed] to the axil of a twig or a leaf or into a petiole.

5. Frequently the bacterial streaks run in the wood close to the pith, which cannot be said of the *Graphium* streaks.

After some practice, therefore, it is in most cases already determinable macroscopically whether one is dealing with a *Graphium* or with a [p. 344] bacterial attack. Microscopically, the bacteria are not always visible in a bacterial discoloration; sometimes they clot together and then can no longer be discerned.

I cannot point out enough that through these bacteria serious mistakes can be made with regard to the cause of the elm disease. *Thus many times in the younger branches of trees severely afflicted by Graphium I have found only bacterial streaks.* An uninitiated person who gets [got] hold of such branches could very easily ascribe the elm disease to bacteria. *The study exclusively of young twigs of a diseased elm* therefore leads to error.

I have conducted very many inoculation tests with this bacterium, in order to be able to compare its effect with that of *Graphium*. It turns out, then, that dark streaks, which can extend far, spread upward and downward in the wood from the point of inoculation soon after inoculation with the bacteria. However, they do not extend into the trunk. Bacteria can be recovered in culture from this discolored wood. *However, by means of bacterial inoculations I have never been able to obtain an acute death, as with Graphium infections.*

The leaves of various twigs inoculated with bacteria did change color early in the autumn, whereupon they abscised earlier than the other leaves.

The bacterial disease of the elm must therefore be considered as a separate disease, which ranks with the attacks caused by Nectria, Verticillium, Phomopsis and which is hardly of a serious nature. The description of this bacterial species will be published later.

The cause of the true, epidemic elm disease is Graphium ulmi Schwarz.

SUMMARY [original in German]

By renewed investigations it is established with absolute certainty that the cause of the Dutch elm disease* is the fungus *Graphium ulmi* Schwarz, and on the following grounds:

1. From all twigs afflicted by this elm disease it is possible to obtain the fungus.

2. Artificial infection with *Graphium ulmi* evokes the typical elm disease within two weeks.

3. Besides the so-called Dutch elm disease there are yet some other diseases to be considered, which can be confused in part with the Dutch elm disease. These are the following diseases, which however stand far behind the *Graphium* disease in economic significance: *Nectria cinnabarina*, *Verticillium dahliae*, a *Phomopsis* species, and a hitherto unknown bacterium (not the *Micrococcus* of Brussoff).

*[This may be the earliest occasion on which a *Dutch* author called this the "Dutch" elm disease.]

Ceratostomella ulmi, the Sexual Form of *Graphium ulmi*

Buisman, C. J.

1932. *Ceratostomella ulmi, de geslachtelijke vorm van Graphium ulmi Schwarz.*
Tijdschrift over Plantenziekten 38(1):1-5 & pl. I-III.
[Netherlands Journal of Plant Pathology].
[Also:] Comitē Inzake Bestudeering en Bestrijding van de Iepenziekte, Mededeeling
nr. 7: pages 1-5, plates I-III.
[Committee for the Study and Control of the Elm Disease, Communication No. 7]*

[p. 1] Of *Graphium ulmi* Schwarz, the fungus which causes the elm disease, no sexual form was known until now. This was not so strange, since we do not know the higher fruiting form of a large group of fungi, the "fungi imperfecti."

Various species of the genus *Graphium*, however, have an ascus form belonging to the genus *Ceratostomella*. The connection between these two genera was first discovered by Münch (5), who proved that *Ceratostomella piceae* and *Graphium penicillioides* are different stages of one and the same fungus. McCallum (4) and Lagerberg, Lundberg and Melin (3) confirmed the research of Münch. Münch found a similar connection between *Ceratostomella cana* and a *Graphium* form. Georgevitch (1,2) described a *Ceratostomella, C. quercus*, which likewise possesses a *Graphium* fruiting stage.

The most recent research related to this matter is that of Rumbold (6). She records the connection between a *Ceratostomella* that occurs in America and a *Graphium*. Probably *C. piceae* is involved here, too. She made 23 single-ascospore cultures of this fungus. In all these cultures she obtained the *Graphium* form but perithecia originated in only 4 of the 23. However, when she combined various ones of the 19 other cultures, then sometimes perithecia were indeed formed. Here we have, therefore, a fact that points to the existence of so-called + and − races in this *Ceratostomella* species. Such + and − races are known from many fungus species. They are, as it were, [equivalent to] the two sexes of one species, which together produce the sexual fruiting body.

[p. 2] If we examine these examples of the connection that exists between some *Graphium* species and species of the genus *Ceratostomella*, then it is understandable that Wollenweber (7) conjectured, that with *Graphium ulmi* some *Ceratostomella* should [serve] as the higher fruiting form. Until now, however, it has not yet been possible to demonstrate this.

Since I suspected that in *Graphium ulmi* we also were dealing with + and − races, which in combination should be able to produce an ascus stage, some time ago I had already combined various pure cultures of *Graphium*

*[Translated by F. W. Holmes, 12/16/1962]

ulmi, by uniting two isolates at a time in a test tube with cherry agar. This experiment had no result whatsoever, as no perithecia appeared.

On May 6 of this year, however, I inoculated some twigs of various elms with *Graphium ulmi*, using as inoculum spores belonging to two different isolates. On May 13 I cut off a couple of these twigs and attempted to reisolate the fungus at some distance from the point of inoculation (the experiments had as a goal, among others, to obtain some data about the speed of growth of *Graphium ulmi*). For making these isolations, sterilized pieces of infected twigs were laid out in dishes of cherry agar. Certain of these dishes, in which *Graphium ulmi* developed, were kept for some time. When I wanted to clear away one of them on July 30, I saw on the twig piece (*not* on the agar), very small black specks. Upon microscopic investigation it appeared that these black specks were perithecia which evidently belonged to the genus *Ceratostomella*. Naturally this was not yet in any sense a proof that this *Ceratostomella* and *Graphium ulmi* were different stages of one and the same fungus, but still it gave an indication in which direction a higher fruiting form of *Graphium ulmi* perhaps must be sought.

From twigs of two other trees, which likewise were inoculated on May 6 with spores of *Graphium ulmi* belonging to two different isolates, reisolations thereupon were also made, as well as from a second twig from the tree which had produced *Ceratostomella* next to *Graphium ulmi* in the just-mentioned reisolation. On July 30 these twig pieces were laid out and on August 25, August 29 and September 15 *Ceratostomella* perithecia were established on a twig piece, originating, respectively, from each of the three trees. It could not be assumed that in all three trees *Ceratostomella* should occur secondarily, and since in none of the very many isolations which I have made in the course of the elm disease research have perithecia ever appeared, it must surely be assumed that [p. 3] these perithecia had originated through infection with two races of *Graphium ulmi*.

Thereupon I began to cultivate my *Graphium* pure cultures in various combinations on sterilized elm twigs. It appeared best for this purpose to combine spore suspensions of two isolates at a time, to shake, and put a drop of this suspension on the sterilized elm twig. *In this way it was possible to obtain perithecia in various instances after a few weeks. It should be emphasized that the pure cultures used never yielded perithecia when grown separately.*

To date it has not happened that when isolate A forms perithecia in combination with isolate B and with isolate C, that B and C together also can form perithecia. We can therefore distinguish two groups which are named + and − in the mycological manner of usage.

There is still no indication that these may be distinguishable visually.

It may be further noted that the sclerotia of *Graphium ulmi* (fig. 1) described by Wollenweber and also observed by me, possibly are perithecia not fully developed.

It appears that in the cultures in which perithecia occur these are not always equally numerous. In some cultures they are present in great numbers; in others one finds but few. Münch, McCallum and Lagerberg also state that in some cultures of *Ceratostomella piceae* no perithecia occur. This happens also with the coremial (=*Graphium*) stage. Münch has already established that various single-spore cultures of *Ceratostomella piceae* possess a highly varying ability to make coremia.

If we compare our *Ceratostomella* with those described earlier, then it appears from the size of the perithecium and the length of the neck that it differs from all the other species. Only *Ceratostomella piceae* Münch has a perithecium of approximately the same size, but this species has a very short neck. We can therefore take our *Ceratostomella* to be an as yet undescribed species.

There follows a description of the *Ceratostomella* which can be obtained by combination of pure cultures of *Graphium ulmi*.

Perithecia black, round, normal diameter 105–135 microns (average 123 microns), seldom entirely bare, mostly provided with a few hairs. Neck frequently somewhat curved, 265–380 microns long, thickness at the base 24–38 microns, above 10–16 microns. At the top of the neck a whorl of cilia, about 24 in [p. 4] number and of very variable length (usually 25–60 microns). Cilia septate and frequently swollen at the tip. The asci are enclosed in a slime mass. The ripe asci disintegrate immediately. Therefore it is not possible to give the size of the asci. Also the number of spores per ascus could not be determined with certainty; probably it amounts to 8. Ascospores slightly curved; [their] shape [is] like that of parts of an orange, size 4.5–6 × 1.5 microns.

The question arises whether this *Ceratostomella* [these perithecia] can play a role from a practical point of view, so far as combatting the elm disease is concerned. It should be emphasized again that the perithecia have not as yet been found in nature but exclusively through combination of various pure cultures of *Graphium ulmi*. The first thing that we must find out, therefore, is whether the perithecia can also be found in nature. For this purpose it is necessary to get a survey of the occurrence of the + and − races, because it is clear that we have no chance of finding perithecia in a region where only one of the two [races] occurs. An investigation into this matter is already under way. Only after the perithecia have been found in nature can it be determined whether the ascospores are of importance for the infection of healthy trees.

It is perhaps possible to get a better insight into the spread of the elm disease with the aid of the + and − races. Thus it is important to examine whether in England, for example, where they assume that the disease was first introduced as late as 1926, perhaps only one of the two groups occurs.

I had brought back from America only *Graphium* material that originated from one sick elm. (This culture made perithecia after it was combined with a culture originating from Baarn). We therefore do not yet know whether both groups are to be found there.

Regarding the greater or lesser pathogenicity of both groups no judgement can yet be offered. The various cultures with which infection tests were in the main carried out this year, belonged by chance to the same group.

Phytopathologisch Laboratorium
"Willie Commelin Scholten," Baarn

Summary

There appear to exist two sexes (so-called + and − races) of *Graphium ulmi* Schwarz. By growing both sexes together on sterilized elm twigs it is possible to obtain perithecia which belong to the genus *Ceratostomella*, and which hereby are described as *Ceratostomella ulmi*.

[p. 5]   Literature List

1. P. Georgevitch. *Ceratostomella quercus* n. sp. Comptes Rendus Acad. des Sciences, vol. 183, 18, pp. 759-76l, 1926. Ref. in Review of Appl. Mycology, vol. 6, p. 198, 1927.
2. P. Georgevitch. *Ceratostomella quercus* n. sp. Ein Parasit der Slawonischen Eichen. Biologia Generalis vol. 3, 3, pp. 245-252, 1927. Ref in Review of Appl. Mycology, vol. 7, p. 206, 1928.
3. F. Lagerberg, G. Lundberg and E. Melin. Biological and Practical Researches into Blueing in Pine and Spruce. Svenska Skogsvardsforenings Tidskrift vol. 2, pp. 145-272, 561-739, 1927.
4. B. D. MacCallum. Some Wood Staining Fungi. The British Mycological Society Transactions, vol. 7, pp. 231-236, 1922.
5. E. Münch. Die Blaufäule des Nadelholzes. Naturwissenschaftl. Zeitschr. für Land- und Forstwirtschaft, vol 5, pp. 531-573, 1907, and vol. 6, pp. 32-47, 297-323, 1908.
6. Caroline T. Rumbold. The Relationship between the Bluestaining Fungi *Ceratostomella* and *Graphium*. Mycologia, vol. 22, pp. 175-179, 1930.
7. H. W. Wollenweber and C. Stapp. Untersuchungen über die als Ulmensterben bekannte Baumkrankheit. Arb. aus der Biol. Reichsanst. vol. 16, pp. 283-324, 1929.
8. F. Zach. Zur Kenntnis von *Ceratostomella pini* Münch. Zeitschrift für Pflanzenkrankh. und Pflanzenschutz, vol. 37, pp. 257-260, 1927.
9. F. Zach. *Über Ceratostomella cana* E. Münch als Varietät von *Ceratostomella piceae* E. Münch. Zeitschrift für Pflanzenkrankh. und Pflanzenschutz, vol. 39, pp. 29-35, 1929.

Summary [original in English]

"The species *Graphium ulmi* Schwarz consists of two sexually different groups (+ and − strains). By growing both sexes together on sterilized elm twigs perithecia can be produced that belong to the genus *Ceratostomella*. A description of this *Ceratostomella* (*Ceratostomella ulmi*) follows.

"Perithecia black, round, normally 105–135 microns (mean 123 microns) in diameter, rarely bare, mostly provided with a few scattered hairs. Necks often slightly curved, 265–380 microns long, breadth at the base 24–38 microns, at the top 10–16 microns. The hyphae [that] the neck consists of diverge at the top and form a crown of cilia. Cilia about 24 in number, septate, sometimes swollen at the top, very variable in length, usually 25–60 microns long. Asci imbedded in a slimy mass. Ripe asci disintegrate in water, which prevents taking their exact size and determining the number of spores per ascus with certainty. Probably an ascus contains 8 spores. Ascospores slightly curved, shaped like the parts of an orange, measuring 4.5–6 × 1.5 microns.

"Perithecia not yet found in nature. They can be produced by combining + and − strains of *Graphium ulmi* Schwarz."

PLATE I

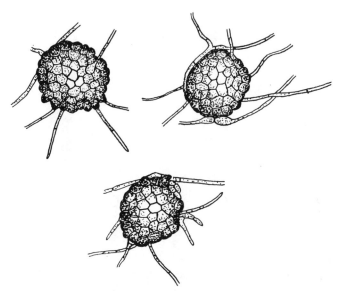

Fig. 1. *Graphium ulmi* Schwarz. Sclerotia (×575).

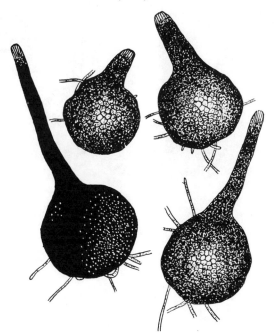

Fig. 2. *Ceratostomella ulmi* (Schwarz) Buisman. Young perithecia (×245).

PLATE II

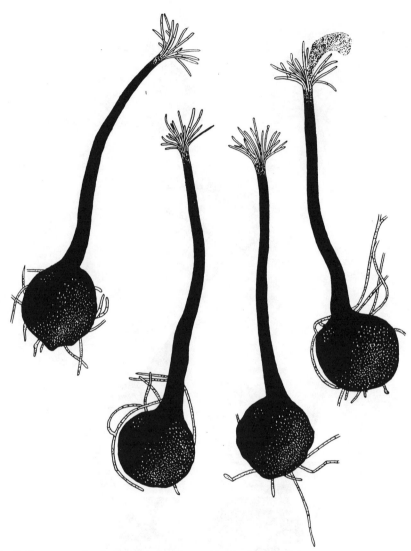

Fig. 3. *Ceratostomella ulmi* (Schwarz) Buisman. Perithecia (×245).

PLATE III

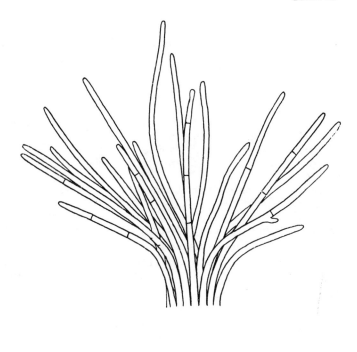

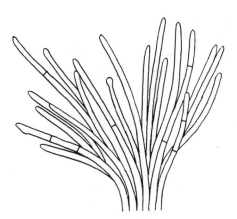

Fig. 4. *Ceratostomella ulmi* (Schwarz) Buisman. Crown of cilia on the top of the neck of the perithecium (×1180).

Fig. 5. *Ceratostomella ulmi* (Schwarz) Buisman. Asco-spores (×1180).

On the Occurrence of *Ceratostomella ulmi*
(Schwarz) Buisman in Nature

Buisman, C. J.
1932. *Over het voorkomen van Ceratostomella ulmi (Schwarz) Buisman in de natuur.*
Tijdschrift over Plantenziekten 38(9):203-204.
[Netherlands Journal of Plant Pathology]
[Also:] Comité Inzake Bestudeering en Bestrijding van de Iepenziekte, Mededeeling
nr. 13:36-37.
[Committee for the Study and Control of the Elm Disease, Communication number
13: pages 36-37]*

[p. 36] A few months ago (Tijdschrift over Plantenziekten, volume 38, January 1932) I described an ascomycete, *Ceratostomella ulmi*, which constitutes the perfect stage of *Graphium ulmi*. The perithecia were obtained through the combination of + and − races of *Graphium ulmi*. The question then immediately arose, whether one would be able to find these perithecia also in the wild.

For answering this question, it had to be determined whether the + and − races were evenly distributed, because only where both occurred can perithecia be expected. In order to determine this we isolated *Graphium ulmi* from twigs originating from trees affected by elm disease from various parts of our country (The Hague, Haarlem, Leeuwarden, Arnhem, Winschoten, Breda, Baarn). It appeared then that in all the places named both + and − races occurred. Also when *Graphium ulmi* was isolated from trees that stood close to one another (Praamgracht, Baarn), we obtained both + and − races.

Since both races therefore appeared to be spread throughout, perithecia could also be expected on every tree, so to speak. It was obvious [only reasonable] to seek them where the coremia of *Graphium ulmi* also occur, namely in the bark of very sick trees attacked by bark beetles. Therefore we collected several pieces of bark from a group of felled elms at Utrecht (on the 3rd of this past March) and kept them in the greenhouse in a very moist environment. *Graphium ulmi* was present in this bark, because this fungus was isolated in large amounts from frass resulting from the galleries of the elm bark beetles. After the pieces of bark had lain for some days in the greenhouse, many coremia of *Graphium ulmi* appeared. On March 16 mature perithecia of *Ceratostomella ulmi* were found on some bark pieces.

Later, a few additional, similar tests were made. On March 3, bark from a sick elm at Baarn was gathered and, just like the bark from Utrecht, kept

*[Translated by F. W. Holmes, 12/16/1962]

moist in a warm [p. 37] place. On April 5, coremia were visible; on April 7, young perithecia—not yet entirely mature.

On April 18, we gathered bark from seven very sick trees at Hedel and this bark was treated in the same way as the previous ones. On April 25, perithecia occurred on the bark of six of the seven trees, mostly still young but in a few cases already mature. The bark of the seventh tree did not produce any coremia of *Graphium ulmi* either; possibly this fungus did not occur in it.

On April 25 we gathered bark from a sick elm from the neighborhood of Baarn. On April 26 this bark was placed in a moist environment in the greenhouse, and on April 29, young perithecia were present, not yet wholly mature but nevertheless already well developed.

From these data it appears that in nearly all investigated cases perithecia were formed in the bark of very sick elms that contained bark beetles. One must assume that in nature, too, whenever temperature and humidity are favorable, not only coremia of *Graphium ulmi* but also perithecia occur in the bark of sick elms. The ascospores can be expected to be transported by the elm bark beetles, just like the coremiospores.

Dr. Marie Ledeboer (left, holding two flasks) in the Dutch National Fungus Culture Collection with Prof. Dr. Westerdijk (center, holding up a flask) and Dr. Harmanna Diddens (right). The photo was taken in 1933 on Prof. Westerdijk's 50th birthday.

CHAPTER 6

Maria Sara Johanna Ledeboer

1904–1988

Maria Sara Johanna Ledeboer was born on December 9, 1904, in Rotterdam, where she spent her childhood. She had one sister and one brother. Both her mother, Petronella Ledeboer-Delprat, and her father, Paul Ledeboer (who had a business in the grain trade), had also been born and reared in Rotterdam.

In 1922, after completing her secondary education at a Rotterdam "gymnasium" (the Dutch equivalent of an academic high school in the United States, with emphasis on Latin and Greek languages and cultures), 17-year-old Maria Ledeboer went to the University of Utrecht to study biology, especially botany. She received her "doctoraal" (equivalent to a Master's degree in the humanities in the United States) there in 1929. There, too, she first met Prof. Johanna Westerdijk.

Under the guidance of Westerdijk, she performed research studies beginning in 1930 in Baarn at the Willie Commelin Scholten Phytopathological Laboratory, and on this basis was awarded her doctorate from the University of Utrecht. On July 2, 1934, she defended her dissertation on the physiology of the Dutch elm disease pathogen, *Graphium ulmi*. During the period of her thesis research, this fungus had first become known also as *Ceratostomella ulmi* (Buisman, 1932).

Dr. Ledeboer's work was a forerunner of extensive studies on the physiology of many species of fungi, as carried out later in other places, notably at West Virginia University by Prof. Barnett and his students. It also foreshadowed an often-felt need, that DED research should turn back to fundamental plant physiology for renewal.

On August 17, 1938, Dr. Ledeboer went with an excursion to South Africa. She ended by staying there for eight years in the Rijksdienst (State Service), ranking as senior pathologist of the Botanical Station at Pietermaritzburg in Cape State. Her responsibilities comprised the whole gamut of tree diseases there, including, for example, studies of diseases of mimosa and of acacia. Among other duties, she had to diagnose specimens sent to the station by citizens in general and especially by tree industries, and to make recommendations for tree disease control.

At the end of World War II, in 1946, when Dr. Ledeboer was 41, she returned to the newly liberated Netherlands. On September 23, 1946, she married H. H. Engel, professor of zoology at the University of Amsterdam. After they had enjoyed a marriage of 37 years, part of the time living in Almen, in the province of Guelder, her husband died in 1983. Dr. Maria Engel-Ledeboer lived for several years in a retirement home for the elderly at Lochem, near Zutphen, northeast of Arnhem (in eastern Netherlands) until her death on August 6, 1988, at the age of 83 years 8 months. She left Ledeboer and Engel relatives in at least six Dutch cities and towns, including stepchildren, grandchildren, and great-grandchildren. Some three and a half years before her death, one of us (Holmes) was fortunate to be given an opportunity to visit her and to talk with her about her South African career, thanks to her step-son, Boudewijn Engel, of the Rijksinstituut voor Volksgezondheid (National Institute of Health) in Bilthoven.

Physiological Research on *Ceratostomella ulmi* (Schwarz) Buisman

Ledeboer, Maria S. J.
1934. *Physiologische onderzoekingen over Ceratostomella ulmi (Schwarz) Buisman.*
[Doctor's dissertation, Univ. of Utrecht, Hollandia-drukkerij: 95 pp. Hollandia Press]*

[p. 1]

INTRODUCTION

In the course of the years various investigations have been carried out on the physiology of fungi. Originally, species from the genera *Penicillium, Aspergillus, Botrytis*, and *Mucor* were principally considered for this; of these, in recent years *Aspergillus niger* van Tieghem has attracted the greatest attention, so that now many physiological features of this fungus are known, while, on the other hand, there are many points of controversy as a logical consequence of the large number of researchers, who—using different methods—often seem to be finding contradictory results.

Thus in this area, seen from the physiological standpoint there is still much that is important to investigate; the results of such experiments gain in general-biological significance if they are confirmed in other test subjects, or rather if they, through particular causes, work out differently. Therefore it is important to compare an investigation into the behavior of a typical parasite with that of *Aspergillus niger*. A fungus which develops primarily in the sapstream of a living tree, as *Ceratostomella ulmi* (Schwarz) Buisman, [and] spreads primarily in the living wood vessels and during the growing season has, in fact, a life-style that totally differs from that of a saprophyte like *Aspergillus niger*.

Not just for comparison with *Aspergillus niger*, but primarily from a phyto-pathological point of view, *Ceratostomella ulmi* was chosen as experimental subject for these investigations: for, as it might turn out that the instigator of the elm disease showed a certain sensitivity toward certain elements or materials which the elm was able to endure, there would exist the greater chance to eradicate the disease. Some knowledge [p. 2] of the physiology of the fungus was indispensible for that research, as one needs to know the secondary effect of the addition of any particular substance if one wants to be able to judge its specific effects.

---

*[Translated by F. W. Holmes, 8/5/1983. This translation contains only the Introduction, pp. 1–2; Conclusions, pp. 87–88; and Summary, pp. 89–91.]

In this connection, the one publication that poses this very problem must be named, to wit, that of Boudru (1933); this [publication] appeared when my research was nearly completed. Boudru studied the fungicidal effect of different dyes with a method which was not able to give a definitive answer about the more general physiology of *Ceratostomella ulmi*; the few preliminary experiments that he did about this were repeated by me and yielded results other than those on which Boudru based his conclusions.

The instigator [cause] of the elm disease is called here by its correct name *Ceratostomella ulmi* (Schwarz) Buisman, despite [the fact that] for experimental material the "Graphium" form was used, i.e., the fungus as it is isolated from the diseased elm wood, thus not from the combination of a + and a − strain that is required for obtaining perithecia of *Ceratostomella*.

[p. 87]   CONCLUSIONS

Possible Control Methods for *Ceratostomella* *ulmi* with Chemical Substances

As was already argued in the introduction, the chance exists that an investigation into the sensitivities of *Ceratostomella ulmi* [might] yield results for direct control of the elm disease. At the same time, emphasis was laid there on the fact that such an investigation can lead to conclusions only if the salient points of the physiology of the fungus are already known. Therefore the goal of this research was, in the first place, to establish these points, before the influence of fungicides was pursued. Since, as is also evident in the literature, most poisons in very low concentrations stimulate various organisms, it turned out that by looking into the stimulative effects of particular elements in the medium, known poisons came into consideration and, on the other hand, materials which were investigated for their toxicity to *Ceratostomella ulmi* turned out in great dilution to be favorable ingredient[s] of the nutrient solution.

In reviewing the possibility which exists to combat the elm disease on the basis of the results of this research, one must take into consideration that the direct control of the parasite will not comprise simple spraying of the diseased (or healthy) trees, since the materials then will not reach the fungus. As a result one must consider instead some internal therapy, e.g., by injection of the fungicidal material into the vessels of the elm. Therefore any material to be applied which would very greatly injure *Ceratostomella ulmi*, may not, on the other hand, in any case injure the elm. So the results to be anticipated cannot be judged before tests are undertaken in this direction.

[p. 88] No positive indications for a suitable material are to be derived from my experiments. To be sure, most elements in a certain concentration retarded the growth, but that concentration is so high that an application of it within the elm is unthinkable. $CuSO_4$ for example, a material which undoubtedly is strongly poisonous to the elm, in my tests was very injurious to *Ceratosto-*

124

mella ulmi in concentrations of no less than 0.01%; of $HgCl_2$ 0.001% was still tolerated; also 0.01% thereof is needed to put *Ceratostomella ulmi* out of action. Lower concentrations of both materials stimulate, so from that point of view, too, applications of them would be risky.

In this direction, however, there still exist many possibilities. Undertaking experiments with the method I used, the use of other fungicidal materials, such as the dyes whose effects Boudru investigated in agar media, deserves to be recommended.

[p. 89]   SUMMARY [originally in German]

The physiology of *Ceratostomella ulmi* (Schwarz) Buisman, the cause of the famous elm disease, was studied primarily in synthetic nutrient solutions. In order to get a high yield, it appeared necessary to start with young cultures which already had grown in synthetic nutrient solution. In order to follow the course of the growth, the yield was often weighed.

1. *Influence of temperature and light*. Even at 8.5° C the growth was already rather good. The optimum lay at 25°, the maximum at about 34° C. Especially that part of the sunlight which was not absorbed by ordinary glass promoted the development of coremia.

2. *Influence of pH*. Each week the pH of some cultures was determined and thereafter the pH of the remaining ones was restored to the original level. The minimum of the routine nutrient solution lay at about 5, the optimum between 6 and 7, and the maximum at about 8. The bounds of the pH seemed to be dependent on the composition of the nutrient solution. With the presence of peptone in the solution, growth was still possible at a pH of 3.6. Peptone, however, occupies a special position with regard to pH, in contrast to other nitrogen sources, for with peptone the pH can be raised during the growth, while with other N-compounds the solution becomes continually more acid and only after the beginning of autolysis will react more alkalinely. The strongest acid level that was determined was a pH of 2.4.

3. *Nutrients*. Not only is the relationship of the nutrients to each other important, but equally [important is] their total concentration.

[p. 90] C—As the carbohydrate source, mostly 5% saccharose was used, as a considerable yield was thus obtained after only 3 weeks. In 5% glucose the maximal yield was approximately the same; but in the first 3 weeks cultures with this carbohydrate lagged behind those with saccharose. *Ceratostomella ulmi* is able to utilize also the following materials as carbohydrate sources: maltose, lactose, galactose, glycerin, mannitol, and starch, of which 5% always exerted a more favorable effect than 1%. Palmitol can be split under exceptional conditions. Cellulose is not attacked; peptone is also unusable as a carbohydrate source.

N—NH_4-compounds as well as asparagine and peptone can function as N-sources. Yet urea is a bad N-source and KNO_3 is totally unusable.

Further, the influence of different metal ions on the growth was studied.

K—is of very great influence, if it is supplied to the nutrient solution as 0.1% KH_2PO_4 or as 0.001%–0.01% K_2SO_4.

Na—NaCl is injurious in a concentration of 0.1%. Yet it was not clear whether the injurious effect arose from Na or Cl. An antagonistic effect between Na and Cl was not discernible.

Mg—is indispensible. A reasonably good growth first arose with the presence of 0.0015% $MgSO_4$. The optimal dosage was 0.15%.

Ca—has a favorable influence on the growth, especially if the Mg concentration lies between the above-named minimum and optimum. Antagonistic effects between Mg and Ca were not found.

Zn—has shown itself to be as important for the structure of the mycelial layer. Without Zn supply, it is more transparent and looser than when 0.002%–0.02% $ZnSO_4$ occurs in the solution. Mostly, 0.1% $ZnSO_4$ limits the growth clearly.

Fe—does not belong to the essential metals. For an antagonistic effect against Zn, at best indications exist.

Cu—Limited Cu concentrations (depending on the composition of the nutrient solution) can be stimulatory to [p. 91] the growth. Higher concentrations are limiting to growth, or prevent it. The fungus unmistakably shows an adaptability to high concentrations.

Mn—A stimulating effect by $MnCO_3$ and $MnSO_4$ was observed. Here, too, the stimulating effect is dependent on the rest of the nutrient solution.

Hg—Low concentrations of $HgCl_2$ stimulated *Ceratostomella ulmi*. The maximal concentration lies between 0.001% and 0.01% $HgCl_2$.

4. *Tannin.* Stimulatory effects by tannin were not observable. The maximal concentration amounted to 1%. *Ceratostomella ulmi* forms a clear halo [areola, corona] on beerwort-tannin agar.

5. *Growth-enhancing materials.* The composition of the nutrient solution can be so altered during the growth, that the pH minimum is lowered. A growth-promoting substance that might cause this could not be detected. With "Bios" (Wassink) [reference to a paper published in 1934; Bios was a kind of yeast extract] the growth was at first strongly accelerated, but through the appearance of autolysis the maximal yield lagged behind the controls.

6. *Control possibilities by chemical means.* The investigations carried out here have offered no basis for this, since the limiting [inhibitory] concentrations lie too high to be taken into consideration for the control of the elm disease, in that, with the materials to be used, a very strong injury would appear in the elms, too. Instead, other materials that are inhibiting to *Ceratostomella ulmi*—ones which were not investigated here—could be effective in combatting the elm disease.

CHAPTER 7

Johanna Catharina Went

1905-

Johanna Catharina Went, called "Hanneke" by her friends, was born on June 14, 1905, in Utrecht. She has one brother, Fritz, who later became professor of plant physiology at the University of California. Their mother, the former Catharine Tonckens, was born in the city of Groningen, capital of the northeasternmost province of The Netherlands. Her father, Frederick A. F. C. Went, who was born in Amsterdam, was professor of Botany at the University of Utrecht.

For her first 30 years, Johanna Went lived in Utrecht. She completed her doctor's degree at the University of Utrecht on June 11, 1934, in the field of phytopathology, under Prof. Johanna Westerdijk, with a dissertation on *Fusarium* infections of peas.

From 1930 to 1934 Went helped Dr. Christine Buisman at the Willie Commelin Scholten Phytopathological Laboratory in Baarn. In 1936 she carried out postdoctoral studies in the United States on tobacco mosaic virus with Prof. James Johnson at the University of Wisconsin in Madison.

Upon the sudden death of Dr. Buisman in 1936, Dr. Went was appointed to carry on her work in selection and breeding of elms for resistance to the Dutch elm disease (DED). This research was sponsored by the Iepenziekte Comite (Elm Disease Committee) for the next 10 years, until 1946, and by TNO (Toegepast Natuurwetenschappelijk Onderzoek, i.e., Applied Research in the Natural Sciences) from 1946 until 1953.

In 1953 Dr. Went moved on, into soil microbiology, as an employee of ITBON (now known as the Rijksinstituut voor Natuurbeheer, i.e., Research Institute for Nature Management). At that time, the elm breeding was put into the hands of Ir. Hans M. Heybroek and a few years later transferred administratively to the Forestry Experiment Station "De Dorschkamp" in Wageningen.

But it was more than a decade before all of the elm breeding work could be moved, since so many important parent trees for the breeding were still located in Baarn. So for a good many years, Heybroek was stationed in the Willie Commelin Scholten Laboratory, even though employed by De Dorschkamp.

Dr. Johanna West (face toward camera) with her assistant, Ms. D. van Wessem on a double ladder, pollinating elm trees on a cold day in early spring 1951 in the Groeneveld research nursery, Baarn.

In DED circles, Went is noted for the fact that she successfully contrived to continue unabated the elm breeding program for disease resistance throughout the World War II years, right under the noses of the Nazi occupiers. The breadth and scope of her research from the very first moment of her shouldering this burden are shown in the accompanying sample of her annual reports. The report from the first year was chosen to illustrate the transition from Buisman to Went.

Dr. Went retired from ITBON in 1970. She now lives in Arnhem, only a few miles east of the Agricultural University at Wageningen, along the north bank of the Rhine River.

Report on the Investigations Concerning the Elm Disease, Carried out at the Willie Commelin Scholten Phytopathological Laboratory at Baarn, During 1936

Went, Johanna C.

1937. *Verslag van de onderzoekingen betreffende de iepenziekte, verricht op het Phytopathologisch Laboratorium "Willie Commelin Scholten" te Baarn, gedurende 1936.* Tijdschrift over Plantenziekten 63(4):75-90, 9 tables. (H. Veenman & Zonen, Wageningen); [also published by:] Iepenziektecomite Mededeeling 24:1-16.

[Netherlands Journal of Plant Pathology. Also: Elm Disease Committee Communication]*

INTRODUCTION

The research into the susceptibility of different elm species and seedlings was carried forward as a continuation of the research of Dr. BUISMAN. Mr. P.J. BELLS acted as seasonal collaborator [assistant].

Most emphasis was laid on the research into the susceptibility of different selected numbers of *Ulmus foliacea* [=*U. carpinifolia*] and *U. glabra*. Besides these inoculation experiments, carried out in the usual way, a few numbers and species were further inoculated with the aid of beetles.

Also the spread of *Ceratostomella ulmi* in elm branches was investigated, in horizontal [lateral] as well as in vertical direction.

The experiments on the influence of nutrition on the inoculation were continued.

SPREAD OF ELM DISEASE IN EUROPE

The area in which it is known that *Ceratostomella ulmi* occurs as parasite of the elm still expands annually.

In February of 1936 Dr. JOSE BENITO MARTINEZ sent material from Burgos (Spain), from which it turns out that this country, too, can now be included with the countries where the elm disease occurs. This was verified by isolation of the fungus. In a communication of the Instituto Forestal de Investigaciones y Experiencias, La Moncloa, Madrid, volume 9 number 15, Martinez describes this find.

From a report of Dr. PEACE, Forestry Station, Oxford, done [p. 76] as an assignment of the Forestry Commission, it turns out that the elm disease

*[Translated by F. W. Holmes, 3/1985]

is still spreading strongly in England also. Cases have now become known from the North of England; not as yet, however, from Scotland.

[p. 77] INOCULATIONS

Because of the long and late period of cold weather in April and May, the elms came slowly into leaf. It seemed desirable, therefore, to begin inoculating as late as the beginning of June.

The strong rainfall during the summer had a noticeable influence on the degree of affliction, so that many individuals were less strongly afflicted than was to be expected. In addition, checking at the end of July and August involved many difficulties, in that many branches were damaged and broken as a consequence of the windstorm at the end of July.

INOCULATIONS OF EUROPEAN ELM SPECIES

The inoculations of *Ulmus foliacea* were continued in relation to [those of] the previous year.

The 3- and 4-year-old Spanish seedlings, which were received from Spain in 1935 and then survived very poorly as a result of the drought, were inoculated this year, although the 3-year-old seedlings grew badly this year, too. A new shipment of Spanish seedlings, which was expected in the spring of 1936, has not arrived.

Inoculations were further carried out on *Ulmus glabra* and *Ulmus glabra fastigiata* also.

From Colmar (Elzas), through mediation of Mr. SELARIES, budwood was received this spring from variant types of *Ulmus glabra* from the Vogezen, which hitherto had remained free of elm disease despite their standing between affected trees.

The results of these inoculations are given in tables I and II.

The result of the inoculations of seedlings in table I shows again that a large proportion of the [*Ulmus*] *foliacea* seedlings is susceptible. The remaining [surviving] seedlings will be further selected some following year.

The inoculations this year of *Ulmus glabra* (Colmar), see table II, give no certainty as yet about the susceptibility, since 1-year-old grafts are much less susceptible than older grafts. It was already possible, however, to eliminate a few susceptible numbers.

INOCULATION OF SELECTED SEEDLINGS
AND THE GRAFTS CULTIVATED FROM THEM

The inoculations were repeated with [tree clones with] the same numbers as were investigated the previous year. This year two new numbers were added, to wit number 52 and number 43. Budwood of these was supplied to the city nurseries of The Hague, Haarlem, and Utrecht and the nursery of the Ministry of Public Works at Amersfoort.

Number 52 is a seedling of *Ulmus foliacea* whose seed originated from France; number 43, a seedling of *Ulmus foliacea* the seed of which came from Spain.

The results of the least susceptible species and of numbers 52 and 43 are provided extensively in tables III and IV; of the more susceptible species, the results are provided in table V in a somewhat condensed way.

From table III it appears that number 24 [later released as the Christine Buisman elm], also, has held up excellently again this year. On the mother tree, one of the 7 inoculated branches showed wilting symptoms. The foliage fell off from a side branchlet of this branch; otherwise the tree grew on normally. Very many grafts of number 24 were tossed about by the storm at the end of July. Much injury came about thereby, so that in a number of cases it was impossible to draw conclusions.

As is to be seen further from the table, *Ulmus foliacea* numbers 28, 31, 42, and 44, and *Ulmus glabra* number 49 have held up well this year. Of these, number 44 is definitely the least suitable form. With this number, even in the control trees, much yellow foliage consistently occurs and in a few places a severe infection by *Nectria* arose.

In table V the results of the susceptible numbers are given. If one compares these with the results of the previous year, then it clearly appears that the inoculations have given less result. This is quite explicable if one compares the weather conditions of the two years; 1935 had a dry summer, while in 1936, in contrast, very much precipitation fell in the summer.

In numbers 27, 29, 35, 39, 46, 47, and 48 no additional inoculations will be carried out, since in the course of 4 years (numbers 46, 47, and 48 only 3 years) these numbers kept showing a very severe infection. The individuals that became diseased mostly appeared to be very severely affected and showed no satisfactory recovery.

In the inoculated [trees of the] numbers [used] in Amersfoort, the severity of the discoloration of the vessels was studied. It turned out that number 24, at a distance of 5 cm under the inoculation point, still showed only a [p. 78] most limited discoloration. This same was the case with number 42, while number 31 showed a somewhat stronger (however, still very limited) discoloration. With numbers 28 and 44, however, a much stronger discoloration arose, which was comparable to that of the more susceptible numbers 26, 29, 33, 39, 46, and 48. Only number 47, where practically no discoloration arose, did not fit here. This number, however, stood in a worse spot in the nursery and thereby had grown very slowly. This fits again with the observation that badly growing trees are much less susceptible. Not enough individuals could be sacrificed, however, to study whether this correlation of severity of discoloration with infection grade consistently occurred.

132

INOCULATIONS OF ASIAN ELM SPECIES

The inoculations of *Ulmus Wallichiana*, of which [tree] a large number of 1-year-old grafts was present this year, were repeated. The seedlings of *Ulmus pumila* that had originated from the nursery of Ansaloni in Italy, were also reinoculated this year.

Root cuttings of a corky *Ulmus pumila* from Texas, which were propagated this spring, did not root fast enough to be inoculated as early as this year.

In Table VI, beside the species named above, a few other Asiatic species are also mentioned.

The older graft of *Ulmus Wallichiana*, which was inoculated, turned out to become diseased, while the one-year-old grafts showed no disease symptoms. However, it has been mentioned above that this last is no criterion for the unsusceptibility of the species.

The *Ulmus pumila* of Ansaloni (Bologna) showed no disease symptoms. Some individuals, however, dropped their foliage very early in the autumn.

The results of the inoculations on the selected seedlings of *Ulmus pumila* are presented in Table VII. As had been reported earlier, these are presumably hybrids of *Ulmus pumila* [i.e., interspecific-hybrid seedlings of which *U. pumila* was one of the parents (and *U. hollandica* 'Belgica' presumably the other)].

Ulmus pumila C showed a strong discoloration early in the year and died off in many cases. This phenomenon occurred in Amersfoort, Haarlem, and Utrecht and is presumably to be explained by a poor connection between understock and scion. *Ulmus pumila* A showed these same symptoms, but in far less degree. This made it very difficult to draw conclusions from the inoculations that were carried out on [trees of] these numbers. *Ulmus pumila* A and B held up well in general; however, occasional disease cases arose just the same.

[p. 79] The inoculations of *Ulmus pumila* D, E, and F took place only on 1-year-old grafts and thus give no definitive answer about the susceptibilities.

INOCULATIONS WITH AID OF ELM BARK BEETLES

At the end of the report of Dr. BUISMAN about the inoculation with the aid of elm bark beetles in 1935, she presented the desirability of repeating these experiments with numbers 24 and 44 and *Ulmus Wallichiana*.

Insofar as this was possible, these inoculations were carried out this year. The method used is described in the report of FRANSEN and BUISMAN, 1935. The beetles originated from a dead tree which had remained standing near the town of Eemnes. The inoculations on number 24, 44 and *Ulmus Wallichiana* took place between [June] 3 and June 18.

Of 9 individuals of *Ulmus foliacea* number 24, after 4 or 5 days, 8, 10, 10, 11, 11, 13, 16, and 16 beetles were recaptured alive and, respectively, 10, 24, 26, 24, 13, 20, 14, and 19 feeding spots were counted. In none of

133

these cases did disease symptoms occur, nor did they on the 5 controls (inoculated with an injection syringe [hypodermic needle]).

From 4 individuals of *Ulmus foliacea* number 44, 4, 8, 9, and 10 beetles were recaptured alive and, respectively, 11, 15, 10, and 13 feeding spots were counted. Here, too, no disease symptoms arose; however, they did with one of the two trees inoculated with an injection needle.

Only two individuals of *Ulmus Wallichiana* could be inoculated with beetles. Here 9 and 10 beetles were recovered alive and 18 and 21 feeding spots were counted. Here, too, no disease symptoms occurred. The single control became diseased.

The results of the inoculation with beetles on these three species went negatively. However, since the inoculation experiment of *Ulmus americana*, carried out with beetles, as [a] control, also went negatively, no additional conclusions can be drawn from this. It is therefore highly desirable to repeat these experiments in some later year.

PENETRATION OF *CERATOSTOMELLA ULMI* INTO THE FOLLOWING ANNUAL RING (HORIZONTAL DIRECTION)

In 1935 special attention was bestowed upon the penetration of *Ceratostomella ulmi* in a horizontal direction with different seedlings, which had been inoculated the previous year. It appeared desirable this year [1936] too, to gather data about this.

Of 49 individuals of *Ulmus glabra*, which [p. 80] were inoculated on June 15, 1935, [it] was determined microscopically whether penetration occurred into the following annual ring. In one *Ulmus glabra*, wilt symptoms arose as early as May 28, 1936, thus without inoculation of the tree in 1936. A strong discoloration was found in the wood of 1936. This same effect was seen on July 1, on another seedling of *Ulmus glabra*, while on this date penetration also occurred into the wood of 1936 in 4 additional seedlings; however, in these last 4 individuals, without outward disease symptoms. Also, on July 24 discoloration was found in the new wood of yet one more *Ulmus glabra* and here, too, no outward disease symptoms arose. In none of these cases did inoculation take place in 1936.

Whereas in the previous year, of the 98 individuals studied, 7 showed penetration into the new wood as well as wilting symptoms, this year penetration by *Ceratostomella* into the new wood occurred in 7 of the 49 individuals of *Ulmus glabra* studied, but only in two cases did wilt symptoms also occur.

Of *Ulmus foliacea*, this year only 11 individuals were investigated for the penetration of *Ceratostomella*. These elms were last inoculated in 1934. In one case penetration occurred into the wood of 1935 and in two cases penetration into the wood of 1935 and 1936. In neither of these last two cases, however, did wilt symptoms occur.

In 1935, 91 *Ulmus foliacea* were investigated without finding penetration in a single case. The penetration thus occurred to a greater degree this year

[1936] than in the preceding year, although the weather conditions for the occurrence of the disease were more favorable in the preceding year.

EXPERIMENTS ON THE SPEED WITH WHICH *CERATOSTOMELLA ULMI* EXTENDS ITSELF LONGITUDINALLY IN ELM WOOD

In 1935 the speed of the spread of *Ceratostomella ulmi* above the point of inoculation in twigs of different elm species was studied by Dr. BUISMAN. This year the penetration *below* the inoculation point was looked into.

For this, seedlings of *Ulmus glabra* and *Ulmus americana* and branches of trees of *Ulmus pumila* were used. Since seedlings were worked with primarily, the material was far from uniform, which shows distinctly in the results.

In order to make comparison possible with the results of the previous year, work was done with the same time intervals, so that 2, 4, and 6 days after the inoculation attempts were made to isolate *Ceratostomella* [p. 81] from the branches at different distances below the inoculation point.

Most isolations were carried out with *Ulmus glabra* and *Ulmus americana*, thus principally with seedlings. These seedlings were mostly too small to inoculate different side branches. For each set of isolations, therefore, another plant was used consistently, which in part can explain the differences that occurred in the isolations. Naturally, these differences also were influenced by other factors, such as variation in external conditions.

The circumstance that in many cases *Ceratostomella* alternately is and is not isolated from consecutive portions of a branch causes difficulty in the evaluation of the results. In a few cases this probably can be ascribed to the method used. For if the branch got too thick to be split in the plating out in Petri dishes, only splinters of the branches are taken off for isolation. If the fungus spreads one-sidedly [only on one side of the branch], it is understandable that one isolates *Ceratostomella* in one case and does not in the other. It is also possible, of course, that the disinfection of a branch has penetrated too strongly, so that the fungus was killed. This, however, is not very probable since all pieces of a branch are sterilized simultaneously and hence under precisely the same conditions.

This, however, does not explain the fact that in some cases *Ceratostomella* was isolated [at points] up to some distance from the inoculation place but not from a next successive piece, and yet *Ceratostomella* could again be isolated from the base. This phenomenon occurs in Table VIII, with *Ulmus glabra* a, b, c, k, and n, and with *Ulmus americana* e and f. The elm material used appeared in several cases to be [have been] already infected a previous year. It now is conceivable, in the cases where *Ceratostomella* occurred close to the base, that an infection had already taken place in a previous year; however, nothing was to be seen of it in the cut-off piece.

In the consideration of the results, these few points—where *Ceratostomella* suddenly showed up at a great distance from the previous point—are disregarded.

If the results, given in table VIII, are compared with the data of the previous year, it then is evident that with *Ulmus americana* a further spread took place above than below the point of inoculation. After two days *Ceratostomella* was isolated at a distance of 30, 40 and 70 cm above and 0, 10, 20, 30, 40 and 50 cm below the inoculation. After four days these distances were: upward 40, 100, and 110 cm; downward: [p. 82] 0, 20, 20, and 20 cm. In *Ulmus pumila* very little difference appeared. *Ulmus glabra* was not used last year for growth-extent experiments; therefore the upward growth-extent also was measured with this species this year. These results, however, presumably do not represent the correct distance of the growth extension above the inoculation place, since the branches were not long enough to make isolation possible at greater distance. In Table IX, the results of the growth extension are stated per branch both above and below the point of inoculation. From this it also clearly appears that the growth extension goes faster above the inoculation place than below the inoculation place. The difference in growth extension below the inoculation place after 2, 4, and 6 days is small.

When one compares the growth extension with *Ulmus glabra* in the different inoculation periods, then it is obvious that the growth extension at the beginning of July is much faster than [that] at the end of May. If, in connection with this, one compares the temperature during this time, then the results strongly indicate a faster growth extension at high temperature, in that high temperatures occurred in the second half of June. From June 17 to 24 the maximum in DeBilt was above 25°C. Also the precipitation in June remained below normal. Thus these are highly favorable factors for the infection to proceed rapidly. On the other hand, the beginning of June was cold. The temperature was 3° to 4° below normal. [A difference of 3–4° Celsius means a difference of 5.4–7.2° Fahrenheit.]

INVESTIGATION INTO THE INFLUENCE OF NUTRITION ON THE COURSE OF THE INOCULATION*

The results of the investigation into the influence of nutrition on the inoculation [infection] remain just as negative as the previous year.

The same series of plants was used this year [1936] as the previous year [1935]. The advantage of this was that the plants were better rooted in the pot, and thereby perhaps would be more susceptible. However, these elms in pots continued to show growth that was not entirely normal.

As routine feeding, a solution of VAN DER CRONE was poured on the pots. During the summer, [a solution containing] 2.5 g KNO_3, 1.25 g $MgSO_4$, 1.25 g $CaSO_4$, 0.625 g $Fe_3(PO_4)_2$, and 0.625 g $Ca_3(PO_4)_2$ was supplied. It

_____

*[She probably meant "infection."]

turned out, however, that in the pots where an additional 2.5 g $Ca(NO_3)_2$ and 2 g $NaNO_3$ was added, leading to a more liberal N-nutrition, a much better growth occurred than in the pots with only salts in the proportion of VAN DER CRONE. These plants (with richer N-nutrition) showed a normal green color, which did not occur elsewhere in this experiment. All other pots thus have a too-restricted N-nutrition.

[p. 83] The plants with little Ca (VAN DER CRONE with 1.755 g $MgSO_4$, 0.126 g $CaSO_4$ and 0.125 g $Ca_3(PO_4)_2$) almost all had dead terminal buds. The plants with much K (VAN DER CRONE with 2.5 g K_2SO_4 and 2.5 g K_2HPO_4) lost all their foliage very early, while the plants with little K (VAN DER CRONE with 0.25 g KNO_3, 0.625 g NH_3NO_4 and 0.625 g $NaNO_3$ were the very ones that stayed green [for a] very long [time].

The rest of the plants, with little N (VAN DER CRONE with 0.25 g KNO_3, 1.0 g K_2SO_4 and 1.0 g KH_2PO_4), much $CaCO_3$ (VAN DER CRONE without $Ca_3(PO_4)_2$ and $CaSO_4$, but with 5 g $CaCO_3$ and 0.625 g $Ca(H_2PO_4)_2$) and much $CaSO_4$ (VAN DER CRONE with 6 g $CaSO_4$) showed little deviation from the pots that got only [the standard] VAN DER CRONE. Six pots were used for each nutrient [treatment]. But of the pots with VAN DER CRONE and with little K, a double number was present. Of these, the previous year 4 and 8 pots, respectively, had been inoculated. This year half of these inoculated were inoculated anew, while half of the controls of the previous year were also inoculated. The other half was used as control.

With one of the controls with abundant N-nutrition of this year, which had been inoculated last year, very beautiful disease symptoms arose spontaneously. This was a beautiful case of the recurrence of the infection of one year in the following year.

In none of the inoculated plants did symptoms arise that were comparable to this. There was not a single case here where one could speak of clear disease symptoms. Only in comparison with the controls was any difference visible. A clear effect on the infection appeared with low K, while at high K there was no effect at all. Some influence was noticeable with high and low N and high $CaCO_3$. The remaining pots [other treatments] showed little or scarcely any effect on the infection.

For comparison, [there] follow, below, the results of the previous year, observed on the same plants with the same nutrition. Disease symptoms seemed then to arise with normal nutrition (VAN DER CRONE), low K, low Ca, high $CaSO_4$ and high $CaCO_3$); on the other hand, not with abundant N-nutrition and high K.

Thus there does not exist very much agreement between the results of these last two years. The only agreement is the limited effect which, upon infection, the plants with high K show, and the distinct effect on the plants with low K.

Consequently it is important to set out experiments on the influence of the K-nutrition on infection in a more extensive form next year.

TABLE I.
INOCULATIONS OF SEEDLINGS OF EUROPEAN ELM SPECIES[1]

Species	Origin	Location	1st inoculation			2nd inoculation			3rd inoculation		
			No.	+	−	No.	+	−	No.	+	−
U. foliacea	France	Baarn	25	15	10	10	5	5	5	-	5
			56	9	47	47	5	42	42	1	41
		Utrecht	86	24	62	60	1	59			
		Bakkum	7	1	6						
	Romania	Baarn	10	4	6	6	4	2	2	1	1
	Greece	Baarn	34	7	27	27	5	22	22	-	22
	Spain, 3 yr	Utrecht	165	13	152						
	Spain, 4 yr	Utrecht	86	49	37	37	4	33			
U. glabra	England	Baarn	120	94	26	26	14	12	12	-	12
U. glabra fastigiata	Baarn	Baarn	29	2	27	27	11	16	16	6	10

[1]The total number of inoculations is indicated by "No.," the number of positive inoculations (diseased) by +, and the number of negative inoculations (healthy) by −.

TABLE II.
INOCULATIONS OF GRAFTS OF EUROPEAN ELM SPECIES[1]

Species	Location	Description	1st inoculation			2nd inoculation		
			No.	+	−	No.	+	−
U. glabra fastigiata	Amersfoort	3-yr graft	10	5	5	5	1	4
U. glabra Colmar 1	Den Haag	1-yr graft	1	1	-	[no further		
2			3	-	3	data in this		
[No. 3 missing]						column]		
4			5	1+3?	1			
5			1	-	1			
6			3	1?	2			
7			3	-	3			
8			3	-	3			
9			3	-	3			
10			3	-	3			
11			3	-	3			

[1]The total number of inoculations is indicated by "No.," the number of positive inoculations (diseased) by +, and the number of negative inoculations (healthy) by −.

TABLE III.
INOCULATIONS OF THE MOST RESISTANT SELECTED [CLONE] NUMBERS[1]

No.	Location	Description	Inoculation date [day-month]	Number of inoculated branches or individuals	Number positive
24	Baarn	3-yr graft	3-6	2	0
24	Baarn	M	3-6	1	0
24	Baarn	M	5-6	4	1 branch slightly diseased
24	Baarn	3-yr graft	5-6	1	0
24	Baarn	3-yr graft	6-6	2	0
24	Amersfoort	3-yr graft	12-6	1	0
24	Utrecht	older graft	15-6	19	0
24	Utrecht	older graft	15-6	20	0
24	Utrecht	2-yr graft	16-6	6	0
24	Haarlem	2-yr graft	16-6	2	0
24	Haarlem	4-yr graft	16-6	7	0 (2 single yellow leaves; 2 broken off)
24	Den Haag	older graft	18-6	1	0
24	Amersfoort	older graft	18-6	10	0
24	Den Haag-S.	layers	22-6	10	0
24	Baarn	M	30-6	2	0
24	Utrecht	older graft	3-7	10	0 (2 broken off by storm)
24	Utrecht	older graft	11-7	19	0
24	Utrecht	older graft	11-7	20	0
24	Utrecht	2-yr graft	17-7	6	0
24	Amersfoort	older graft	20-7	10	0
24	Amersfoort	1-yr graft	21-7	15	0
24	Haarlem	4-yr graft	24-7	8	0
24	Den Haag-S.	layers	27-7	15	0
24	Den Hagg-S.	?-year graft	27-7	11	0
28	Utrecht	M	9-6	3	0 (1 with yellow leaves)
28	Amersfoort	3-yr graft	12-6	3	0
28	Amersfoort	2-yr graft	12-6	9	1
28	Haarlem	2-yr graft	16-6	3	1 slightly diseased (1 some leaves fallen off)
28	Utrecht	2-yr graft	16-6	10	0
28	Utrecht	M	3-7	2	0
28	Utrecht	2-yr graft	17-7	10	1 very slightly diseased (1 much foliage lost)
28	Amersfoort	older graft	20-7	3	0
28	Amersfoort	1-yr graft	20-7	7	0 (stagnant growth)
31	Utrecht	M	3-6	2	2 very slightly diseased
31	Den Haag	3-yr graft	11-6	2	1 slightly diseased
31	Amersfoort	3-yr graft	12-6	8	0
31	Amersfoort	older graft	18-6	1	0
31	Utrecht	M	3-7	2	0

[continued on next page]

[TABLE III continued]

No.	Location	Description	Inoculation date [day-month]	Number of inoculated branches or individuals	Number positive
31	Den Haag	3-yr graft	11-7	1	0
31	Amersfoort	3-yr graft	20-7	8	0
31	Amersfoort	older graft	20-7	1	0
31	Amersfoort	1-yr graft	21-7	2	0
31	Den Haag	1-yr graft	25-7	4	0
[p. 86]					
42	Amersfoort	3-yr graft	12-6	10	0
42	Haarlem	3-yr graft	16-6	3	0 (1 stagnated)
42	Amersfoort	older graft	18-6	1	0
42	Baarn	M	20-6	2	0
42	Amersfoort	3-yr graft	20-7	10	0 (1 slightly stagnated)
42	Amersfoort	older graft	20-7	1	0
42	Haarlem	3-yr graft	24-7	4	0
42	Baarn	M	29-7	2	0
44	Baarn	M	4-6	4	1 slightly diseased (many yellow leaves)
44	Den Haag	3-yr graft	11-6	5	0 (3 some yellow leaves)
44	Amersfoort	3-yr graft	12-6	7	1
44	Haarlem	3-yr graft	16-6	4	0 (1 growth stagnation, 2 some yellow leaves, 1 many yellow leaves, *Nectria*)
44	Den Haag	3-yr graft	18-6	1	0
44	Baarn	M	30-6	2	0
44	Den Haag	3-yr graft	10-7	4	0 (2 leaves with brown edges)
44	Amersfoort	3-yr graft	20-7	6	0
44	Haarlem	3-yr graft	24-7	1	0
49	Den Haag	2-yr graft	11-6	3	0 (2 with light shoots)
49	Baarn	M	11-6	1	0
49	Haarlem	2-yr graft	16-6	1	0 (1 a few leaves becoming brown)
49	Utrecht	2-yr graft	16-6	6	0
49	Den Haag	2-yr graft	11-7	3	0
49	Utrecht	2-yr graft	17-7	6	0
49	Haarlem	2-yr graft	24-7	1	0 (1 some yellow leaves)
49	Baarn	M	29-7	2	0

[1] M = mother tree [= ortet].

[on p. 86] TABLE IV.
INOCULATIONS ON THE NUMBERS NEWLY GRAFTED IN 1936

No.	Location	Description	Inoculation date [day-month]	Number of inoculated branches or individuals	Number positive
43	Baarn	M	20-6	3	0
43	The Hague	1-yr graft	22-7	4	0
43	Utrecht	1-yr graft	23-7	1	0
43	Baarn	M	29-7	2	0
52	Baarn	M	4-6	4	1 branch diseased and top diseased
52	Amersfoort	1-yr graft	21-7	7	0
52	The Hague	1-yr graft	22-7	4	0
52	Utrecht	1-yr graft	23-7	7	0

[p. 87] TABLE V.
INOCULATIONS ON THE SUSCEPTIBLE NUMBERS OF THE CHOSEN SEEDLINGS

Species and number		Number of inoculated individuals	Number of positive infections
U. foliacea	26	23	3 & 1 slightly diseased
	27	3	2
	29	28	9 & 4 slightly diseased & 3 ?
	35	14	4
	39	25	9 & 1 slightly diseased & 1 ?
	40	3	0
	46	36	6
U. hollandica klemmer	47	20	4 & 1 slightly diseased
U. japonica	48	25	13 & 3 slightly diseased
U. glabra	50	16	7
U. foliacea	51	31	6

TABLE VI.
INOCULATIONS OF ASIATIC ELM SPECIES

Species	Location	Description	Inoculation date [day-month]	Number of inoculated branches or individuals	Number positive
U. pumila	Baarn	seedling Italy	5-6	69	0
	Baarn	seedling Italy	8-6	79	0
	Baarn	seedling Italy	3-7	79	0
	Baarn	seedling Italy	7-7	69	0 (5 lose their foliage too early in the fall)
	The Hague	2-yr graft	18-6	5	0
	The Hague	2-yr graft	11-7	1	0
U. pumila pendula	The Hague	2-yr graft	18-6	2	0
	The Hague	2-yr graft	11-7	2	0
Karagatch elm	Utrecht	older graft	15-6	4	0
	Haarlem	older graft	16-6	3	1 slightly diseased (2 some brown leaves)
	Utrecht	older graft	11-7	4	0 (2 die back, but control also: probably rootstock [incompatibility]
	Haarlem	older graft	24-7	3	0 (1 much foliage fallen off, brown)
U. Sieboldii	The Hague	2-yr graft	25-7	4	0
U. lac. nikk.	Haarlem	4-yr graft	16-6	1	0
	Haarlem	4-yr graft	24-7	2	0
	The Hague	?-yr graft	25-7	4	1
U. Wallichiana	The Hague	4-yr graft	18-6	1	1
	The Hague	1-yr graft	22-7	30	0 (1 with yellow foliage)
U. macrocarpa	The Hague	2-yr graft	25-7	4	1

TABLE VII.
INOCULATIONS ON THE GRAFTS OF THE SELECTED SEEDLINGS OF *U. PUMILA*

Species	Location	Description	Inoculation date [day-month]	Number of inoculated branches or individuals	Number positive
U. pumila A	Amersfoort	2-yr graft	12-6	10	1 slightly diseased
A	Haarlem	2-yr graft	16-6	3	1 (1 some yellow leaves
A	Utrecht	2-yr graft	17-6	20	0
A	The Hague	3-yr graft	18-6	5	0 (2 some dry leaves; 1 dried margin & leaves)
A	The Hague	3-yr graft	11-7	3	0
A	Utrecht	2-yr graft	17-7	20	2
A	Amersfoort	2-yr graft	20-7	9	1
A	The Hague-S.	?-yr graft	27-7	5	1?
U. pumila B	Amersfoort	2-yr graft	12-6	6	0
B	Utrecht	2-yr graft	16-6	10	1?
B	Haarlem	2-yr graft	16-6	4	0 (2 some yellow leaves)
B	The Hague	3-yr graft	16-6	2	0
B	The Hague	3-yr graft	11-7	2	0
B	Utrecht	2-yr graft	17-7	10	0 (somewhat more yellow foliage)
B	Amersfoort	2-yr graft	20-7	6	2 + 2 slightly diseased (1 dried foliage; 1 red discolored foliage)
B	Haarlem	2-yr graft	24-7	4	0
B	The Hague-S.	?-yr graft	27-7	15	0
U. pumila C	Amersfoort	2-yr graft	12-6	4	1 + 1?
C	Utrecht	2-yr graft	16-6	10	[blank]
C	Haarlem	2-yr graft	16-6	1	[blank]
C	Utrecht	2-yr graft	17-7	10	0
C	Amersfoort	2-yr graft	20-7	3	0
U. pumila D	The Hague	1-yr graft	22-7	25	0
U. pumila E	The Hague	1-yr graft	22-7	13	0 (2 with somewhat dry little leaves)
U. pumila F	The Hague	1-yr graft	22-7	10	0

TABLE VIII.
ISOLATION OF *CERATOSTOMELLA ULMI* AT DIFFERENT DISTANCES UNDER THE INOCULATION PLACE ON DIFFERENT ELM SPECIES

Species	Inoculation date	No.of days after which was isolated	Twig[1] No.	10	15	20	25	30	35	40	45	50	60	70	80	85	90	100	110	120
glabra	25-5	2	a	+	+	−	−					+	+							
			b	+	−	−	+					−	−							
		4	c	+	+	−						−	−	−	+			−	+	
			d	+	−	−						−	−							
		5	e		+					−				−	−		−	−		
			f		+					−				−	−					
			g		+					+										
	6-6	2	h	+	+	+	+	+		+		−								
		4	i		+					−		−	−							
		6	j		+				+	+		+	−							
			k		+					+		−	+	−	+					
	13-6	2	l	+	+	+	+	+	+	+		+	+	−	−			−		
			m	+	+	−	−	−	−	−		−	−	−						
		4	n	+	+	+	−	−	−	−		−	+	−	−			−	−	−
			o	+	−	+	+	−	−			−	−	−						
		6	p	+		+			+		+	+	+	+	+			−	−	−
	23-6	2	r	+	+	+		+	+	+	+	+	−	+						
		6	s		+	−	−	−		+	−	−	−	−			−	−	−	−
	11-7	2	t	+	+	−	+					+	+	+	+			+	+	
		4	u	+	+	+	+					+	+	+	+			+	+	
americana	25-5	2	a	+	+	+	+			+										
		4	b	+	+		−			−		−	−							
	6-6	2	c	−	−	−	−	−		+										
	1-7	2	d	+	+	+	+	+		+		−	−	−	−			−	−	−
		3	e	+	+	+	+	+	+	−		−	−	−	−			−	−	−
	6-7	2	f_1	+	−	−	−	−	−	−		−	+	−	−			−		
			f_2	+	+	+	−	−	−	−		−	−	−	−			−		−
		4	f_3	+	+		−	−				−	−	−	−		−			
	11-7	2	i_1	+	+	+	−					−	−	+	−			+		
		4	i_2	+	+	−	−					−	−	−	−			−	−	
			k	−	−	−	−					−	−	+	−			+	−	
pumila	25-5	2	a	+	+	+	−		−	−										
	6-6	2	b	+	+	−	−	−												
		4	c	+	+	−														
		6	d	+	+	+	+													

[Translated table, i.e., a mirror image of original, rotated 90° to fit page.]

[1] If several twigs have originated from a single plant, they get the same letter, e.g., *U. americana* f_1, f_2, f_3.

TABLE IX.
PENETRATION OF *CERATOSTOMELLA ULMI* ABOVE AND UNDER THE INOCULATION POINT, WITH *ULMUS GLABRA*

Number of days after which was isolated:	2 days			4 days			5 days	6 days	
Number of the twig [branch]:	a	b	h	c	d	i	g	j	k
Growth extent above inoculation place (cm)	40	20	70	40	30	50	120	70	80
Growth extent below inoculation place (cm)	20	10	40	20	20	20	40	50	40

CHAPTER 8

Louise Catharina Petronella Kerling

1900–1985

Louise Catharina Petronella Kerling, known to many as "Loes," was born July 19, 1900, in The Hague. She had one sister, Nelly Johanna Kerling (1910–1978). Their father was Johannes Jacobus Kerling (1868–1927), a physician, and their mother was Johanna Petronella Ebling (1873–1962).

From 1922 to 1926 Kerling was a Botany Department assistant at the University of Leiden. In 1925 she passed her *doctoraal* (somewhat equivalent to an American Master's degree) examination there. For two summer vacations she worked in Wageningen at the Plantenziektenkundige Dienst (PD, i.e., Phytopathological Service, a government agency that included plant disease diagnostic, advisory, and quarantine/regulatory responsibilities). Here she met N. van Poeteren, T. A. C. Schroevers, and Dina Spierenburg at the height of the PD's involvement with the newly discovered Dutch elm disease (DED).

During another summer, Kerling worked for Prof. H. M. Quanjer in the Potato and Mycology Laboratory at the Landbouwhogeschool (Agricultural University) in Wageningen. It followed naturally that, in November 1926, she became an assistant at that University (to Prof. E. Reinders in the Botany Department). She did her doctor's thesis on the anatomical structure of leaf spot diseases under Prof. Johanna Westerdijk, however, and so it was the University of Utrecht that awarded her a doctorate on July 2, 1928.

In June 1929, Kerling went to the Dutch East Indies, to teach in the government secondary school at Jogjakarta. After a year's leave in The Netherlands (1935–1936), she returned, first (1936–1939) to the H. B. S. (university track secondary school) at Medan, Sumatra, and then (1939–1942) again to Jogjakarta.

Like so many of her generation, Kerling found her life strongly affected by both world wars. In 1942 the Jogjakarta school closed under the Japanese occupation. Dr. H. H. Thung helped her to get appointed to the General Agricultural Experiment Station at Buitenzorg, Java, but after only one year there she was interned (for more than two years) in Kamp Halmaheira in Semarang. Here, to help keep up camp morale, she taught and lectured. She survived, eating toads if she could catch them, and studying dysentery at the camp medical laboratory.

Prof. Dr. Louisa C. P. Kerling (standing), asking a question as one of the examiners at the doctor dissertation defense of Gé van den Ende on 12 December 1958 at the Agricultural University in Wageningen.

After the war, while recuperating in Australia, Kerling was guest researcher in the Waite Agricultural Research Institute, Adelaide. She then was invited by Prof. H. M. Quanjer to come back to Wageningen, to become plant pathologist in the Laboratorium voor Fytopathologie in the Agricultural University. She worked there from 1946 until 1952, when she was appointed successor to Prof. Westerdijk as professor of plant pathology at the Universities of Utrecht and Amsterdam and as director of the Willie Commelin Scholten Phytopathological Laboratory in Baarn. These tasks she fulfilled for the next 18 years, until her retirement in 1970.

Kerling's personal contribution to DED research was but a small part of her career. In 1955 she published a meticulous study on the reactions of elm wood to the DED fungus: this paper needs no translation in the present Phytopathological Classic, as it was published in English in *Acta Botanica Neerlandica* 4(3):398-403. In 1968 she published again about DED, this time about its events in America. And the DED thesis studies of Rebel and Elgersma were done under her direction. Moreover, it was Kerling who had an entire corridor of DED scientists, all at once, at the Willie Commelin Scholten Laboratory. And it was she who had the vision to appoint a plant physiologist (Elgersma) to study DED. The excerpts from her "Fifty Years of Elm Disease" included here cover only the earliest years of the DED, but her career coincided with the entire first half-century of that disease. Kerling died at her home in Baarn on August 24, 1985, at the age of 85.

Biographical material may be found in the *Netherlands Journal of Plant Pathology* 76:110-118 (1970) and a memorial, in *Phytopathology* 76(1):39 (1986).

Fifty Years of the Elm Disease

Kerling, L. C. P., and D. M. Elgersma
1970. *Vijftig jaar iepziekte.* Vakblad voor Biologen (december):273-282, 6 figs, 36 refs.
[Newsletter for Biologists. Only the early-history portions are given here: parts of pages 273-277 and 282.]*

FIRST APPEARANCE OF THE DISEASE

The aspect of The Netherlands must indeed have altered drastically by the cutting of thousands of stately elms which, planted along the roads and dikes, conferred upon the landscape a singular beauty. By estimate there were still 1,128,000 [elm] trees to be found along 10,000 kilometers [6,200 miles] of roads in 1930; in 1946, only 425,000 remained [see Fig. 1].

The disease, to which the elms *Ulmus hollandica* 'Belgica' fell sacrifice, apparently claimed the first victims in 1917, but only in 1919 did it get attention, by [its] especially rapid spread through the whole country. The same disease symptoms also showed up in northern France and Belgium.

* * *

THE CAUSE OF THE DISEASE

Dina SPIERENBURG, phytopathologist with the Plant Disease Service, gave reviews of the distribution of the disease in 1921 and 1922. In her efforts to track down its cause, she isolated a number of fungi from the discolored woody parts, among these a *Graphium* sp., with which she carried out inoculation experiments. The typical wilting symptoms, however, did not appear, so that she was inclined to think of a disease of a physiogenic nature, more than [to think] of an infectious disease.

Meanwhile, M. Beatrice SCHWARZ (1922) worked on the same question in Baarn under guidance of Professor Johanna WESTERDIJK. She described *Graphium ulmi* [p. 274] n.sp. as the disease cause, a fungus which can show various fruiting forms upon culture in vitro. After introduction into a healthy elm, the typical local dark discolorations arose in the wood.

According to SCHWARZ the origin of an infection should be sought in the leaves. Only in 1927, however, was wilting of the leaves of a tree observed after inoculation with a conidial suspension of the fungus—by Christine BUISMAN, likewise a student of Johanna WESTERDIJK. She discovered

*[Translated by F. W. Holmes, 7/1985]

Figure 1. Elm-lined road in Friesland. [Tree-tops in foreground show dieback. Some of these trees still remained in 1985.]

that only inoculations carried out in the period between the end of May and beginning of July could result in this effect.

In those years phytopathologists, as well as nonspecialists, were still in no sense ready to regard *G. ulmi* as cause of this disease. But WOLLENWEBER, a recognized authority in the area of plant disease and mycology, independently confirmed the results of SCHWARZ. Johanna WESTERDIJK (1928) not only powerfully defended the research of her earlier student, but also indicated the approach to combatting the disease: the development of resistant elm clones. It was the research of these women that procured for the elm disease the name "Dutch elm disease," for which people abroad often apologize.

* * *

THE PARASITE

SCHWARZ (1922) described the coremia or synnemata, brush-shaped bundles of upright conidiophores, the tips of which bear conidia of the *Graphium* type on small teeth. Conidia can also arise at the tips of swollen hyphae, with sterigma-like outgrowths, a form wrongly referred to as the "*Cephalosporium*-type." Detached spores, as well as those still on the sporophores, can sprout, yeastlike, by which practically no hyphae are formed.

* * *

BUISMAN (1932) discovered the perfect stage of the parasite: perithecia with a long neck, and hair-shaped hyphae around the mouth opening. The walls of the asci turn into slime at maturity and the ascospores come out in a slime drop. The fungus is now generally designated as *Ceratocystis ulmi* (Buisman) C. Moreau. Already Buisman knew that the fungus is heterothallic.

* * *

To the genus *Ceratocystis* belong mostly wood- and bark-inhabiting fungi, many of which cause "blue wood" [blue stain]. Others live together with beetles, which consume their mycelium. The symbiosis of *C. ulmi* with elm bark beetles has led to a strong spread of this fungus. It is unknown how and where this combination came about, and why the disease did not become established earlier in our land, formerly so rich in elms.

THE SELECTING OF ELMS RESISTANT TO *C. ULMI*

One of the aspects of the elm disease question which had the interest of Christine BUISMAN, was the search for a replacement for the 'Belgica' elm, one which is resistant to the disease. A large collection of elms was assembled, in the form of grafts or seed of trees from many lands. Approximately 10,000 seedlings were tested for resistance by injecting a spore suspension into trunks or large branches. By far, the most plants proved to be susceptible.

A seedling originating from Spain turned out [p. 276] to be resistant to the fungus and was distributed to nursery operators under the name 'Christine Buisman' elm, and an equally resistant clone of French derivation was the 'Bea Schwarz' elm. However, this latter was not generally planted out, because of the bad form of the tree, while in the Dutch climate the 'Christine Buisman' elm was severely affected by *Nectria cinnabarina*, the red-lead mold [coralspot fungus], which made large branches die back. In Italy, however, the tree worked out quite satisfactorily.

Yet other difficulties presented themselves, which had to be taken into account in the selecting. Seedlings that had just been transplanted, no matter how susceptible, do not react to inoculation with a spore suspension of *C. ulmi*, a fact that was known already to Christine BUISMAN (1932). They do display disease symptoms in the year following that [year] when they are transplanted—or not until even later. This can lead to a deceitful "pseudo-resistance," which several times has led to disappointments. It has happened that people were about to release a clone, for which they had high expectations, only to find out that the trees were susceptible after all. It has even happened that an already issued clone turned out to be less resistant than people originally thought they could conclude from the results of the tests.

The method applied for testing for resistance, as this was developed by Christine BUISMAN, was taken over after her death in 1936 and further developed by Johanna C. WENT. Immediately after the seed is mature, they let it germinate in seed flats under the influence of light. The seedlings were planted out, in July or August, close to each other in beds, after which they were transplanted anew a year later at distances of one meter. Only [at the] beginning of the fourth year were they ready to react to an inoculation. After being inoculated for five consecutive years, whichever surviving trees had a pleasant form and good growth were grafted [actually, multiplied by grafting many pieces from the same tree as scions onto seedling or clonal rootstock], after which the grafts were yet again investigated for resistance for five years (Went, 1954).

* * *

HYBRIDIZATION

In 1937 a beginning was made with [genetic] crossings. Much research was carried out by WENT on the method of removing the male floral parts and enclosing the flowers for artificial pollination—a cold job, which had to be carried out from February through April in the trees, high up on a ladder. A genetic basis for the hybridization work still had to be developed. By experimental experience *U. carpinifolia* turned out to be a good producer of pollen and the moderately susceptible *U. hollandica* 'Vegeta' a good mother tree.

151

Gradually the general level of resistance climbed, and other important characters that should be considered in selection also got attention. Besides resistance to the red-lead mold [*Nectria*], there had to be resistance against frost, wind, or [and] salt in the ground.

* * *

[p. 281] After a half-century of research much about the elm disease has become known, but how the complicated disease syndrome ultimately comes into existence is a question on which further research will have to shed light.

Notes and Suggestions
for Users of Any Translation

by F. W. Holmes

Anyone who uses a translation should keep in mind the following:

1. A translation is a service, to be used with caution.

2. Errors inevitably occur in translations and in their reproduction, just as well as in writing and in printing of an original publication.

3. Certain expressions in foreign languages cannot be translated directly into expressions with exactly equivalent meanings in English.

4. No one can ever guarantee the accuracy of the original writer's subject matter observations, descriptions and conclusions, or the accuracy of publication of the original article.

5. With that text as a start, then, the translator makes the best possible attempt—as a professional in the same field—to express competently, in the new language, the ideas of the original author.

6. A good mechanical translation is not possible. A good translation can come about only if the translator completely understands the meaning of the text in its context. Translation therefore involves interpretation. Alternative interpretations, of course, cannot always be excluded.

7. There is no single, perfect way of translating. In different eras, and certainly in different languages, people express themselves in different ways and styles. One translator may choose to translate such texts broadly into a smooth present-day language; another may translate in a more literal way, following the author's text, reasoning and style more closely. For scientific papers, the second option seems more appropriate.

Users of translations are urged to take the following precautions:

1. Try consulting the original publication as well as the translation. Photos, drawings, and long tables sometimes are omitted from translations (not in the present Classic unless expressly stated).

2. Treat a translation preferably as a guide to your own study of the word choice of the original author!

3. For vital passages—particularly passages to be quoted—be cautious. Look for any error in the translation. If the original author is alive, write to the original author. Since original authors are usually no longer available, try asking the translator whether a different interpretation of any vital passage in the text might also be possible.

4. Never cite a translation alone. You cannot, of course, "cite" the originally

publication if you had not seen and consulted it. But as a service to your readers, you should always give the location of the original publication. Thus your readers may have a chance to look at it and to draw their own conclusions as to possible shades of meaning intended by the original author.